The Smithsonian Treasury
MINERALS and GEMS

The Smithsonian Treasury
MINERALS and GEMS

John Sampson White
NATIONAL MUSEUM OF NATURAL HISTORY

Special Photography by Chip Clark

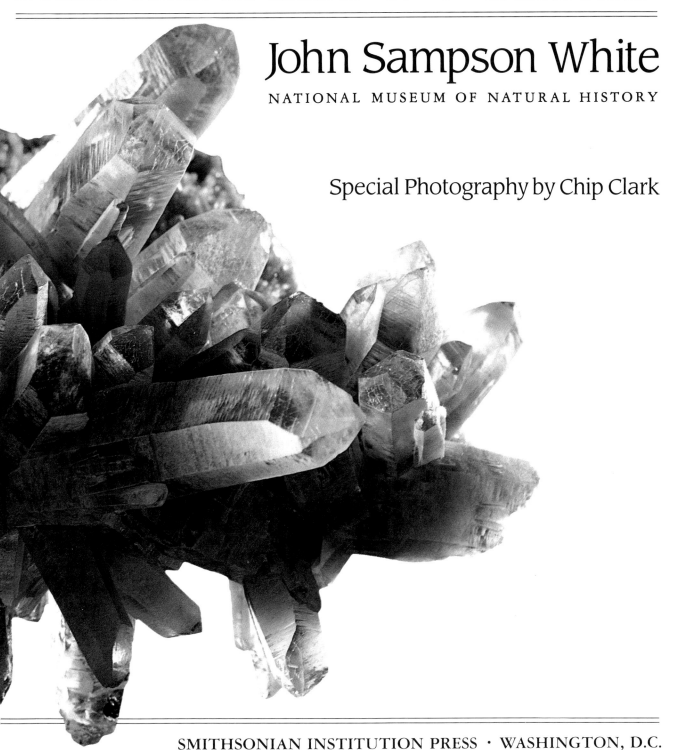

SMITHSONIAN INSTITUTION PRESS · WASHINGTON, D.C.

Copyright © 1991 Smithsonian Institution
All rights reserved

Produced in cooperation with the Book Development Division, Smithsonian Institution Press.

All photographs copyright © Smithsonian Institution, except as noted in the Acknowledgments.
Drawings copyright © Outlet Book Company, Inc.

A Gramercy Book
distributed by Outlet Book Company, Inc.,
a Random House Company,
225 Park Avenue South,
New York, New York 10003.

Manufactured in the United States of America

Library of Congress Cataloging-in-Publication Data
White, John Sampson, 1933–
    Minerals and gems / John Sampson White ; special photography by Chip Clark.
        p.   cm. — (The Smithsonian treasury)
    Includes index.
    ISBN 0-517-05951-7
    1. Mineralogy. 2. Minerals. 3. Gems. I. Clark, Chip.
II. Title. III. Series.
QE363.2.W55   1991
549—dc20   91-17141
                                          CIP

8 7 6 5 4 3 2 1

CONTENTS

FOREWORD 7

The Mineral Kingdom 9

What Is a Mineral? 12

How Minerals Form 16

Crystals 21

Oddities of Crystallization 26

Replacements 30

Physical Properties 32

Classifying Minerals 38

Mineral Groups 40

Modern Mineralogy 42

Collecting Minerals 44	The Garnet Group 72
The National Mineral Collection 46	Jade 74
The World of Gemstones 51	Lesser Known Stones 76
Discovering Gemstones 54	Massive and Decorative Stones 81
How Gemstones Are Cut 56	Organic Gems 84
Diamond 58	Synthetic, Imitation, and Assembled Gems . 86
Corundum 60	The National Gem Collection 88
The Beryl Family 62	INDEX 94
Opal 64	ACKNOWLEDGMENTS 96
Quartz 66	
Topaz 70	

FOREWORD

Minerals, found deep within the earth's crust and upon its surface, have been a source of fascination and delight for thousands of years. Minerals are of great economic significance and are of importance in our everyday lives because, with little modification, they can be applied to a myriad of daily uses. Many minerals are wondrous in their crystalline forms, and some are strikingly beautiful when they are cut and polished as gemstones.

Since 1963, I have been affiliated with the Department of Mineral Sciences at the National Museum of Natural History and actively involved with one of the world's finest collections of minerals and gems. My work has led to extensive contact with people in many fields—other mineralogists, mineral collectors, dealers, teachers, students, politicians, even film stars. But those I have most enjoyed meeting are people who, knowing little or nothing about minerals, have allowed me to share in the extraordinary experience of their first "discovery" of these natural materials.

In this book I have tried to provide a larger audience with a window onto the world of minerals and gems at the Natural History Museum. Here I explain some of the scientific and technical aspects of minerals—their origins, chemistry, physical and optical properties, and the ways they are used. There is a description of how gemstones are cut and information about where and how they are found. You will learn about well-known gems, including diamonds, rubies, and sapphires, many lesser-known stones, and organic gems, as well as the stories behind such famous pieces of jewelry as the Hope diamond and the earrings that once belonged to Marie Antoinette. In addition, the often confusing nomenclature of gemstones is made clearer.

Although some two thousand specimens are on display, these represent only a few percent of the National Museum of Natural History's collection, for most of it was not acquired with exhibition in mind. These minerals are in the collection because of their scientific importance and few of them are even attractive. With this book you have the opportunity to go behind the scenes, glimpse some of the work done there, and learn how thousands of scientists may benefit, directly or indirectly, from this research material.

This book is illustrated throughout with photographs of specimens, almost all of which are part of the national collection. The majority of these magnificent images were taken especially for this book by staff photographer Chip Clark, with whom it has been my great pleasure to work.

I hope that this book will stimulate everyone who reads it to learn more about minerals and gems, which are so diverse and so beautiful that they seldom fail to enchant all who are fortunate enough to encounter them. The mineral and gem displays in the Natural History Museum have made it possible for millions of visitors to have this experience. This book lets the visitor take some of that experience home and relive it and it permits the armchair traveler to visit with each turn of its pages.

JOHN SAMPSON WHITE

This impressive group of emerald-green beryl crystals was found at Hiddenite, North Carolina. It is 4 1/2 inches tall.

The Mineral Kingdom

Everything in the natural world that is not plant or animal, liquid or gas, is mineral. The mineral kingdom encompasses the sands of the world's beaches and deserts, valuable ores mined from deep underground, fantastic stalactites and stalagmites, and rocks of literally every kind. The rocks that make up the world's continents are all made of minerals, as are the rocks at the bottom of the oceans, and the rocks beneath the earth's crust in the planet's mantle. Even the rocks of the moon are made of minerals, most of which are very similar in composition to those of the earth.

About three thousand different minerals have been identified in the earth's crust. Each is a natural substance with a definite chemical composition and specific physical characteristics that distinguish it from other minerals. Rocks are natural masses or aggregates of minerals. Some, like granite, slate, and basalt, are mixtures of several kinds of minerals; others, like marble and limestone, are aggregates of crystals of only one mineral.

Manufactured materials—steel, bronze, brick, plastic, tungsten, phosphorus, soap, plaster, cement—are not minerals because they are not natural in origin. Man-made materials, even if their chemical composition and structure are identical to a mineral's, do not qualify. Quartz crystals and many gems, including ruby, sapphire, and diamond, can be

The Rocky Mountains are a great soaring expanse of minerals massed in different types of rocks.

grown in a laboratory, but these synthetics are not considered true minerals.

Deciding what is or isn't a mineral would be easier if the rules (natural, solid, not animal or vegetable) were rigidly applied, but mineralogy acknowledges a number of exceptions to the rules. Most of these exceptions were named minerals when the descriptive tools for characterizing rocks and minerals were far more primitive. As a result, the mineral kingdom takes in a few natural substances that don't quite fit the definition—coal, oil, and natural or volcanic glass. Although coal is a rock, its crystals are organic substances, often mixed with remnants of plants. Oil is a liquid. Natural glass lacks crystalline structure and will "flow" given enough time. Nevertheless, all three may be considered minerals. Water is also a special case: as naturally frozen ice, it, too, is considered a mineral.

The mineral kingdom is diverse and knowledge of it is still growing. Each year more than fifty "new" minerals are recognized. While this number seems very large, the total known quantity of each of these newly found species seldom exceeds several grams.

Of course, the abundant, more common minerals were discovered long ago. Easily located, they occurred in quantities sufficient to give mineralogists who first described them ample material for tests and analyses. One reason that so many new minerals are turning up today, even though there is so little of each, is that modern analytical equipment requires only microscopic crumbs for precise characterization. Often more can be done now with a microgram of a very scarce mineral than could be done thirty years ago with a crystal the size of a football.

Another reason for the proliferation of new and scarce species is that we are penetrating geological environments that have never before been accessible. Today we mine ores that were impossible to extract or not profitable enough in the past. We also exploit different types of deposits. Technological applications have created demand for previously unused metals—cadmium, hafnium, caesium, rubidium, selenium. Finally, deep drillings have unearthed unknown species here on this planet, while excursions to the moon have led to the discovery of new species in lunar rocks.

QUARTZ

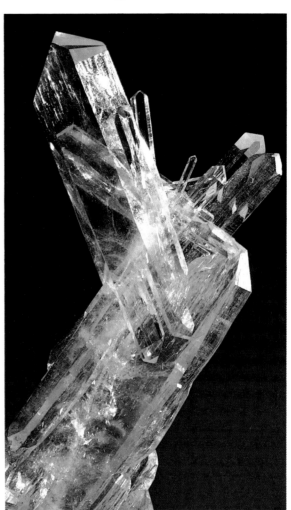

The unofficial flag bearer of the mineral kingdom, quartz is found virtually everywhere—even on the moon, although it is rare there. Many of the rocks that make up the world's great mountains contain quartz, and quartz is the main component of most desert sands—testimony to this mineral's resistance to alteration. Quartz may be found in many other environments in the earth's crust and on its surface. Most of the world's streambeds and riverbeds are carpeted with boulders and sand that are predominantly quartz. Flowing water transports quartz to the ocean's shores, where pounding waves reduce it to sand.

Ubiquitous though it is, quartz is even more remarkable for its showier forms: for it commonly occurs as magnificent crystals like the quartz crystals above.

Quartz is the main component of the sands of deserts and of most of the world's beaches.

What Is a Mineral?

A mineral is a naturally occurring solid substance that has a definite chemical composition and a crystalline atomic structure. *Chemical composition* refers to the elements of which the substance is composed. To take a simple example, common table salt is composed of sodium and chlorine. This familiar chemical compound is also found in nature as the mineral halite. A mineral's crystalline *structure* refers to the orderly spatial arrangement of its atoms. There are only fourteen basic arrangements of regularly repeating patterns of atoms or groups of atoms in three dimensions. Although hundreds of mineral structures have been recognized, all of them conform to one of these fourteen basic symmetrical arrangements.

Chemistry and structure together define a mineral. This is best illustrated by comparing several pairs of minerals. Two or more minerals may have the same chemical composition but different structures. This is the case with pyrite and marcasite, both of which are iron sulfide, and with calcite and aragonite, both calcium carbonate. Conversely, some minerals with identical structures have dissimilar chemical compositions. Quartz and berlinite, for example, have the same structure, but quartz is composed of silicon and oxygen, and berlinite of aluminum and phosphorus.

Structure greatly influences a mineral's physical properties. Apparently quite unlike, diamond and graphite nevertheless have the same composition. Both are carbon, but because of differences in crystal structure, graphite is one of the softest minerals

Although diamond, above left, and graphite, left, have the same chemical composition—carbon—their crystal structures are very different, as shown in the models above. These structural differences are so extreme that diamond is the hardest of all minerals and graphite is one of the softest. (The diamond specimen is from South Africa, the graphite is from Rossie, New York.)

and an excellent lubricant, while diamond is the hardest of all minerals and a prized gemstone when faceted and polished.

A mineral is never "pure." Gold in its natural state is never just gold. It always contains impurities—varying amounts of other metals, of which silver, copper, and iron are only a few.

Some minerals with identical atomic structures belong to a so-called solid solution series, a continuous series of compositional variations. Forsterite and fayalite constitute such a series, formed by varying amounts of magnesium silicate and iron silicate. Any ratio of magnesium to iron is possible in the series. If the mineral contains more magnesium than iron, it is called forsterite; if there is more iron than magnesium, the mineral is called fayalite. Most of the gem variety of forsterite, which is called peridot, is found in volcanic rocks and has a ratio of magnesium to iron of about 6 to 1.

Minerals and Rocks

What distinguishes minerals from rocks? A mineral is a chemical compound with characteristic composition and atomic structure. A rock is formed of minerals, one or more in varying proportions, with millions of crystals bonded together to form a solid mass. Granite and marble, for example, are rocks. Even without a microscope, granite can be seen to consist of interlocking crystals of the minerals quartz and feldspar and, possibly, even a mica-group mineral. The individual crystals in marble are usually much tinier than those in granite. Under magnification, marble reveals interlocking crystals of just one mineral, a carbonate, in most cases either calcite or dolomite.

Rocks are typically found in large masses. A single rock unit may constitute the bulk of an entire mountain range or an entire layer that continues for hundreds of miles. This may be seen most dramatically at the Grand Canyon, where a sequence of rock layers, one on top of another, stretches for miles and miles.

Except for animals, plants, and man-made materials, the solid earth is rock. Most rock is composed of a relatively limited number of minerals, and most of these minerals are silicates.

Silicates are composed of one or more metals, such as iron and magnesium, along with silicon and oxygen. These four elements alone make up approximately 90 percent of the earth. With the addition of the next four most abundant elements—nickel, sulfur, calcium, and aluminum—the total comes close to 99 percent. All the rest of the ele-

Aragonite, above top, and calcite, above, are both calcium carbonate. They have the same chemical composition but different crystal structures. (The aragonite specimen is from Czechoslovakia, the calcite from Tennessee.)

Estimated Chemical Composition of the Earth			
Element	Symbol	Atomic Number	Estimate
Iron	Fe	26	38.3
Oxygen	O	8	28.5
Silicon	Si	14	14.7
Magnesium	Mg	12	7.7
Nickel	Ni	28	3.1
Calcium	Ca	20	2.8
Aluminum	Al	13	2.4
Sodium	Na	11	0.7
Sulfur	S	16	0.7
Titanium	Ti	22	0.3
Chromium	Cr	24	0.2
Potassium	K	19	0.2
Cobalt	Co	27	0.2
Manganese	Mn	25	0.1
Phosphorus	P	15	0.1

The remaining elements total less than 0.1 percent.

WHAT IS A MINERAL?

FOURTEEN BASIC STRUCTURES

Although hundreds of mineral structures have been recognized, all of them conform to one of the fourteen basic symmetrical arrangements of atoms or groups of atoms shown here.

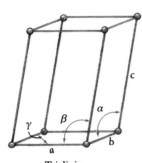
Triclinic
$a \neq b \neq c \neq a$
$\alpha \neq \beta \neq \gamma \neq 90°$

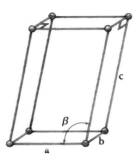
Monoclinic
$a \neq b \neq c \neq a$
$\beta \neq 90°$

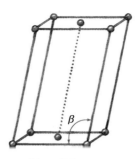
Monoclinic
(end-centered)

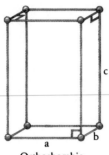
Orthorhombic
$a \neq b \neq c \neq a$

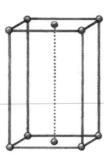
Orthorhombic
(end-centered)

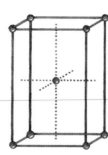
Orthorhombic
(body-centered)

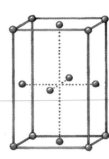
Orthorhombic
(face-centered)

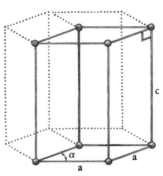
Hexagonal
$a \neq c$
$\alpha = 60°$

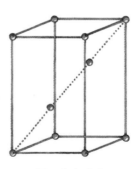
Rhombohedral

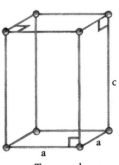
Tetragonal
$a \neq c$

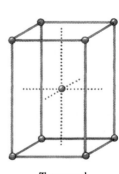
Tetragonal
(body-centered)

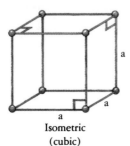
Isometric
(cubic)

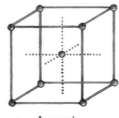
Isometric
(body-centered)

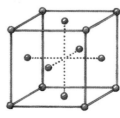
Isometric
(face-centered)

ments combined, some eighty-four of them, constitute less than 2 percent of the earth. Fortunately, over the four and a half billion years of the earth's history, the geological processes that have created and shaped the continents have effectively concentrated many of these elements in deposits that can be mined. Therefore, many elements essential to life as we know it, including copper, iodine, chromium, zinc, lead, and boron, can be recovered from the earth in useful quantities.

Gems

A gemstone can be either a single crystal of a mineral or an aggregate of minerals, including rocks, but it is always the product of shaping and polishing by a gem cutter, or lapidary. Fewer than a hundred mineral species are traditionally used as gems. Some gems, including pearl, coral, amber, and jet, technically are not minerals because they are derived from living organisms.

To be desirable as a gem, a stone must be attractive after it is cut and polished, and sufficiently hard to survive everyday wear when it is set in jewelry or ornamental objects.

Large pieces of sculpture made from gem materials like jade or malachite may be considered gems, but those created from rocks like granite and marble are generally not. The distinction between large ornamental rock carvings, which are not gems, and small stones used for personal adornment, which are gems, is quite arbitrary.

Many people prefer to limit the term "gem" to stones that are usually faceted or cut into small, rounded forms for use in jewelry. But it is important to remember that almost all gems are simply rocks or minerals that have been modified by some combination of sawing, grinding, and polishing.

This exquisite piece is carved of tourmaline, gem-quality elbaite from Mozambique.

A gemstone is always the product of shaping and polishing. Top, from left to right: a 469-carat citrine; a 288-carat kunzite; an 815-carat topaz; and a 12,555-carat topaz. Left, from left to right: a 158-carat synthetic gem; a 121-carat aquamarine; a 202-carat tourmaline cat's-eye; and a 900-carat rose quartz. These stones are all Brazilian, except the 815-carat topaz, which is from the U.S.S.R.

How Minerals Form

Quite naturally, the most detailed knowledge of minerals and rocks and the processes that produce them comes from the earth's crust. All scientists know about deeper rocks in the earth's underlying mantle and core is based entirely on abstract and indirect methods of study—seismic evidence and laboratory studies of rock, coupled with studies of meteorites, which are believed to represent fragments of disintegrated planetary bodies. Much can also be inferred from surface rocks that lay very deep in the earth earlier in their histories.

The size and nature of the mineral crystals that make up the solid earth, and the complexity of the rocks in which they are found, relate very much to the circumstances under which the various rocks formed. The three basic types of rocks—igneous, sedimentary, and metamorphic—all form in different ways.

Igneous Rocks

Igneous rocks crystallize from molten rock, or magma, as it cools. Rapid cooling does not allow much time for orderly growth of crystals, so that when lava bursts from the volcano Kilauea in Hawaii, for example, tiny, often microscopic crystals form. In other cases, molten rocks may cool within the earth's crust, or even far below it, losing heat very slowly, giving crystals more time to grow.

Broad plateaus of black basalt, solidified from massive outpourings of lava, and batholiths, great masses of magma intruded into the earth's crust and later exposed as major mountain ranges through uplift and erosion, are only two products of igneous processes. The minerals that make up the bulk of each of these two dramatic examples, as well as the many other forms of igneous rocks, do

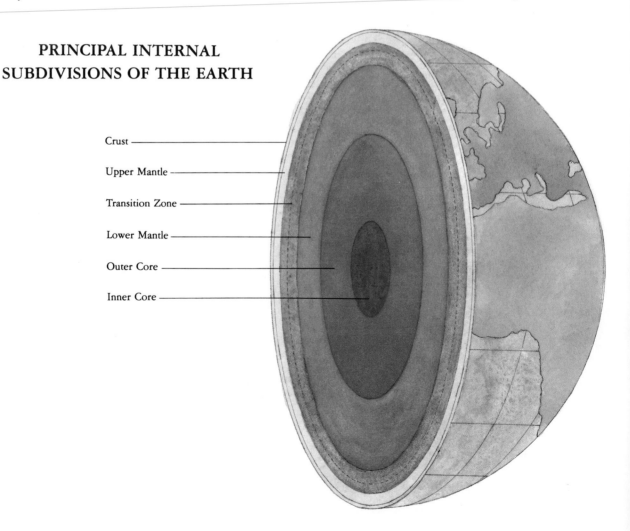

PRINCIPAL INTERNAL
SUBDIVISIONS OF THE EARTH

Crust
Upper Mantle
Transition Zone
Lower Mantle
Outer Core
Inner Core

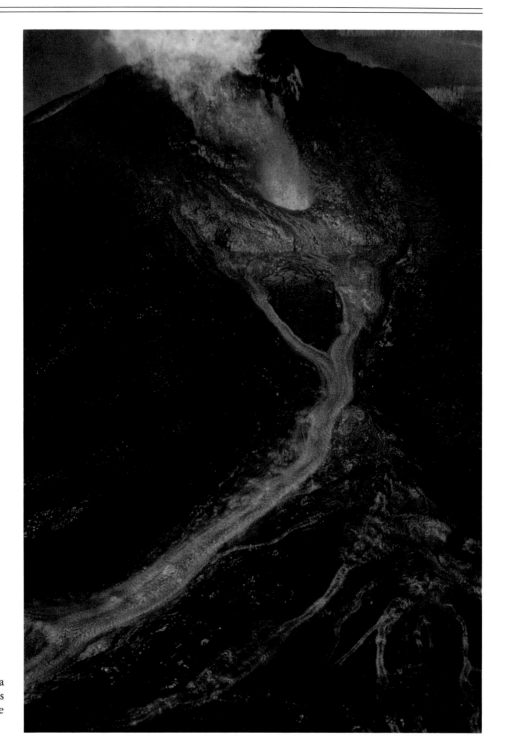

When magma erupts as lava from a volcano, like Hawaii's Kilauea, and then cools at the surface, extrusive rocks form.

not usually produce wondrous specimens of beautiful crystals, because of "overcrowding" during cooling. There are, however, notable exceptions. Rocks formed within fractures in igneous rocks are called pegmatites; some of them are rich sources of many crystals mined for gemstones.

When magmas crystallize deep in the earth, they often give off superheated liquids and gases that collect within the rock or along its edges. In many instances, these fluids are enriched by exotic elements like tin, lithium, beryllium, cesium, and tantalum. Initially thinly dispersed throughout the magma, these elements are not readily incorporated into the crystals growing as the magma cools, and consequently become concentrated in residual solutions. Cooling and crystallization of the magma into a huge mass of rock set up internal stresses that can lead to fracturing. Fractures provide avenues along which residual fluids can move. As they flow away from the main igneous mass, they cool quickly. Crystals begin to grow, now incorporating the remaining exotic elements. If the fluids encounter large openings in rocks, oversized crystals are apt to grow. Uncrowded in the cavities, the crystals can grow from the walls in exquisite forms.

Under special conditions, exciting crystals can form in lava. When basalt, for example, flows out upon the earth's surface and begins to cool, it may release great quantities of dissolved gases. Huge, round bubbles form in the viscous, rapidly crystallizing lava. If the lava continues to flow slowly after the bubbles form, the cavities can stretch, taking on cigarlike shapes. After the lava crystallizes, watery solutions, still probably very hot, may flow through it, picking up silicon as other minerals are dissolved. Upon entering cavities, however, the solutions sometimes precipitate silicon, in combination with oxygen, as quartz crystals. In some cases, minute amounts of iron are also incorporated, producing crystals of purple quartz, or amethyst.

Crystals tend to grow from all the interior walls of a cavity, projecting toward the center. Such crystal-lined hollow objects are known as geodes. The most famous occurrence of amethyst geodes covers many square miles and extends on both sides of the border between southern Brazil and Uruguay.

Minerals may form from vapors, too, usually associated with volcanoes. After the more violent eruptions have ceased, a small number of minerals, particularly sulfur, form directly from gases emitted from fumaroles, the vents around volcanoes.

Sedimentary Rocks

Exposed to a depth of one mile in some places, the Grand Canyon reveals extraordinary examples of rocks that were at one time erosional debris—mostly sand, silt, and gravel. This sediment was deposited by running water, and perhaps by wind, into an ancient sea. From abundant evidence in the canyon's rocks, geologists know that the sea into which the sediment was deposited was not as deep as today's canyon. The sea floor sank at about the same rate as sediment was accumulating on its surface. In time, as a result of compaction and compression, the layers of sediment became consolidated into what are known as sedimentary rocks. Finally, the whole sequence of rock was elevated to its present height above sea level of about 8,000 feet at the top. Vigorous erosion by the Colorado River gouged out a deep gash, exposing spectacular, almost horizontal rock layers. Similar sedimentary deposits are forming today in ocean basins all around the world. A good example is the delta at the mouth of the Mississippi where the river enters the Gulf of Mexico.

In warm seas, like those off the coast of Florida, another type of sedimentary rock is forming. Here the sea water contains large amounts of calcium carbonate, mostly dissolved from the shells of organisms. Under the proper conditions, calcium carbonate will precipitate and accumulate on the sea floor. These accumulations are gradually compressed and hardened into limestone.

Many sedimentary rocks are fine-grained because they are composed of weathered fragments of other rocks, which must be light enough to be carried by water or wind to the sea. At times, however, more energetic environments, such as fast-flowing streams or beaches, may result in the deposition of coarser materials in sedimentary basins, forming conglomerates—rocks containing a mixture of sand and pebbles.

Sedimentary rocks are seldom exciting as mineral specimens because of their fine-grained character, but crystals of a number of minerals grow in

The shell of this rare star geode, from Rio Grande do Sul in Brazil, formed first of milky quartz, then of amethyst, which derives its color from iron impurities. The star pattern resulted when amethyst crystals encrusted earlier calcite crystals.

Almost horizontal, the sedimentary rock layers of the Grand Canyon are spectacular.

openings in the massive rocks, such as fractures or holes left when organisms buried in the sediment subsequently dissolved away. Solutions may flow into these cavities and precipitate minerals. The minerals most often found as crystals in sedimentary rocks include quartz, calcite, aragonite, gypsum, pyrite, and marcasite.

Metamorphic Rocks

Rocks of the third major group have all been changed through recrystallization or chemical alteration of a notably different pre-existing rock. The process of change can take a number of forms.

Many of the rocks of the continents have undergone extreme gyrations over the course of the earth's history. Where mountains stand today, sediment slowly accumulated in ancient basins. Other rocks that now crown high peaks once lay buried deep in the earth. Elsewhere, magma has invaded rocks, pushing them aside or even swallowing vast parts of them. Many of these events exert such profound forces upon the rocks that their components alter in response.

Recrystallization is one common change brought about by great increases in temperature or pressure or both. Adjacent minerals in a rock may react to form new ones. Small crystals may actually melt, then recrystallize as the same minerals but in different forms. Others, especially those that are elongated or plate-shaped, may simply rotate in response to directed pressure, usually the weight of overlying sediments, so that they are all oriented in the same direction. As is the case with igneous and sedimentary rocks, rocks of metamorphic origin

HOW MINERALS FORM

Within schist, a metamorphic rock, staurolite is often found in sharp crystals. This specimen is from Windham, Maine.

are also readily recognized by their distinctive features. To a skilled petrologist these features reveal much of the rocks' histories.

Minerals found in metamorphic rocks are characteristic of the various types of metamorphism, or change, that specific kinds of rocks have experienced. When rocks of different overall chemical compositions are subjected to the same form of metamorphism, they can produce different sets of minerals. Rocks of the same chemical composition, even though they may be composed of quite different minerals, respond similarly to metamorphism; in some instances, identical sets of minerals will be generated in each.

It is possible, for example, for a sedimentary rock and an igneous rock to have almost identical bulk chemical compositions. Under extreme conditions of metamorphism, when the melting points of all the minerals are exceeded and complete recrystallization occurs, the sedimentary and the igneous rocks can become rocks difficult to tell apart.

Granites are good examples of this. Although they are commonly considered igneous in origin, it is evident that many of them started as other types of rock. They have been completely recrystallized and are, therefore, metamorphic, even though they have the texture of igneous rock.

Many metamorphic rocks have within them large crystals that are not only distinctive for the rock type but attractive and desirable as crystal specimens, too. For example, schist, a layered metamorphic rock, commonly contains garnets or staurolite, the latter frequently found in twinned pairs of crystals intergrown at specific angles and referred to as "fairy stones" or "fairy crosses."

When magma invades other rocks, especially carbonates like limestone, the result is somewhat like a smelting furnace. A stewpot of new minerals can form in the reaction between the two. This "contact metamorphism" often leads to the crystallization of zones of many unusual minerals, especially if there are openings in which they can form.

How Fast Minerals Grow

Minerals formed by volcanic processes are among the fastest growers. They can grow in hours or even minutes because lava cools very quickly at the earth's surface. Growth rates of most of the other members of the mineral kingdom remain a mystery. Although equivalents of many of the minerals can be produced in a laboratory in days or weeks, it is not commonly known how long the natural growth period actually is. It is likely that much of the mineral world took hundreds, thousands, or even millions of years to evolve.

THROUGH A PETROGRAPHIC MICROSCOPE

Regardless of whether molten rocks cool quickly and produce small crystals, or slowly so that larger crystals evolve, a mosaic of interlocking crystals is revealed when seen in thin section.

First, a slice of rock is attached to a glass slide and ground down so that the minerals in it become nearly transparent. When this "section" is viewed through a petrographic microscope using polarized light, the minerals can be identified by the colorful optical effects thus created. A thin section also reveals other features of the rock that may tell the petrologist a great deal about the conditions under which it formed.

Crystals

A crystal is a three-dimensional latticework of atoms. Far from being stacked randomly, the atoms interlock in a precise geometrical pattern repeated in all directions throughout the crystal. The fundamental arrangement of the atoms in the basic building block of the crystal, the unit cell, determines the structure of the crystal and defines its symmetry.

With few exceptions, minerals exist as crystals. A crystal grows as atoms attach themselves, layer upon layer, to the surfaces of a seed or baby crystal. If "floating" in another medium, a crystal may grow on all sides so that it becomes larger at approximately the same rate in all directions. If, as is often the case, it is attached to another substance, a crystal may then grow rapidly in only one direction, usually away from that attachment.

An axiom of the crystal world is that smaller crystals tend to be more perfect than larger ones. This is not difficult to understand since a crystal grows as more atoms attach themselves to its surfaces. For a crystal that grows only a millimeter or so each day, these atoms must arrive at the surface, find their niches, and settle in place at the rate of

This cluster of crystals from Naica, in Chihuahua, Mexico, is gypsum, a soft mineral composed of hydrated calcium sulfate.

This is a representation of the crystal structure of halite. The red balls represent chlorine, the blue sodium.

several hundred layers of atoms every second. It is easy, therefore, to understand why "mistakes" occur. Every so often an atom misses its spot and leaves a hole, or an alien atom fills in where it shouldn't be. Such mistakes produce imperfections in the crystals, often causing profound effects not obvious in smaller crystals, but magnified in crystals that have grown very large.

How Crystals Grow

When thousands of crystals are growing simultaneously from a magma that is cooling, they grow into one another on all sides, so that few, if any, can assume the kind of external geometric forms we often associate with crystals. This certainly is the case with most rocks, which are dense masses of tightly interlocking, irregularly shaped crystals resembling a jigsaw puzzle.

The crystals that museums usually display and that mineral hobbyists typically collect differ in appearance from the ones that make up rocks, primarily because most "specimen" crystals evolve differently from rocks.

In the upper reaches of the earth's crust, open spaces may exist within the rocks. These spaces can be gas pockets, open fractures, or solution cavities (including caves) created when liquids dissolved the minerals once present there. At greater depths, because of the pressure of overlying rock, there are no openings. The importance of openings is that they provide space for crystals to grow in a relatively unimpeded fashion. Under favorable conditions, with the right raw materials present, crystals often begin to grow upon the walls of any openings, as with geodes, the hollow rocks lined with crystals. Fluids, usually hot, water-rich solutions, are likely to occupy these openings, and it is out of these fluids that crystals form.

When the fluid contains dissolved silicon, quartz will probably be the mineral to form since its composition is silicon and oxygen. Given sufficient time, quartz appears to crystallize readily under a variety of conditions. The more complex the chemistry of the fluid's composition, the more diverse the kinds of crystals that may form.

The shape of the crystal is greatly affected by such factors as temperature, pressure, and acidity of the solution. Under one set of conditions, the crystals may be hairlike, very long, thin, and delicate. Under other conditions, the same mineral may form stubby crystals. When space permits uninterrupted growth, the resulting crystals are often beautiful and may emerge in elegant groupings.

Crystal Shapes

Several factors control shapes of crystals. To a large extent, the configuration of the unit cell—the basic building block of a crystal—determines a crystal's general external form. Minerals whose atomic arrangement yields unit cells in the shape of a cube, for example, often form crystals whose external

THE SEVEN CRYSTAL SYSTEMS

Crystals are grouped into systems according to their symmetry.

Isometric

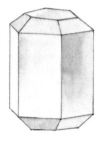
Hexagonal

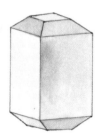
Tetragonal

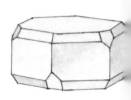
Trigonal

shape is cubic. Other influences, however, may cause a "cubic" crystal to grow at different rates in different directions so that it looks more like a rectangular solid than a cube. In some cases, the crystals of a cubic substance are severely distorted. Experienced collectors become familiar with such distortions and recognize the true symmetry of a crystal even when it is so disguised.

At one time, crystallographers could do little more than carefully measure the relative positions in space of all of a crystal's faces. With just this information alone, it was possible to characterize the crystal's symmetry—a powerful identification tool then. Today, X-ray diffraction can identify the symmetry of almost any crystal's unit cell. All that is needed for a test sample is a tiny fragment, uncontaminated with other minerals.

Through X-ray study, the crystallographer learns the exact position of each atom relative to all others in the basic configuration of the unit cell, even measuring the distance between atoms precisely. Models of the structures can illustrate the results of such an analysis. Small colored spheres represent atoms, which are positioned around one another just as are the atoms of the actual crystal. These models are, however, only representations of the structure and do not resemble what an actual crystal would look like if it were magnified to the scale of the model.

Crystal Systems

Mineralogists organize crystals into seven systems, according to symmetry. The simplest and most symmetrical, the isometric system, is represented by the cube. The other six systems, in order of decreasing symmetry, are hexagonal, tetragonal, trigonal, orthorhombic, monoclinic, and triclinic.

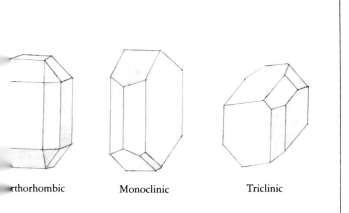

rthorhombic　　　Monoclinic　　　Triclinic

MICROMOUNTS

Of the two thousand or so minerals that form good crystals, most never produce large ones. Consequently, a variation of the mineral hobby has evolved around collecting small specimens containing crystals so tiny that they must be viewed under a binocular microscope. The specimens are usually placed on a tiny pedestal in a small box. The whole arrangement is referred to as a "micromount."

Because specimens of microscopic crystals are generally abundant, the mounts are chosen from the most aesthetic. A skilled micromounter can trim the specimen adeptly and position it on its pedestal artistically so that the viewer sees a truly spectacular dioramalike display of crystals. First-time observers are always stunned by the beauty they see when they put their eyes to the microscope. Not only are the crystals quite perfect, but the colors are usually vivid and the artistry of the composition exquisite.

Wulfenite on torbernite from Shaba, Zaire.

Metatyuyamunite from Shaba, Zaire.

CRYSTAL HABITS

Shape and size give rise to descriptive terms applied to the appearance, or "habit," of crystals.

The crystals of some minerals, for example, usually form in long needles, tapered or pointed at the ends. These crystals are described as having an acicular habit. Mineral crystals that are elongated but do not taper or come to a sharp point are said to be prismatic, particularly if the sides are bounded by long, narrow faces. Other crystals tend to be stubbier and may actually bulge at the middle so they resemble a barrel, and therefore have a barrel-shaped habit. If crystals are long and flat, they are bladed. Extremely long and thin hairlike crystals are filiform; curiously, they are usually flexible, like hair, even though larger crystals of the same mineral are quite brittle.

Aggregates of crystals may adopt forms that are often characteristic of certain minerals. Hematite, for example, is frequently found in colloform or reniform groupings, named for its spheroidal forms that resemble bunches of grapes, or perhaps kidneys. Other frequently used terms for similar features are botryoidal, mammillary, and tuberose. Foliated or micaceous minerals tend to part readily along cleavage planes to form thin sheets. Those that break into fibers are said to have fibrous habits; those that come apart as granules are, of course, granular. A dendritic habit describes minerals whose crystals branch treelike from a central trunk, usually in just two dimensions. More feathery growths are called plumose. Needlelike crystals that converge from many directions to a point form stellate, or star-shaped, groups.

The many terms used by mineralogists for habits are useful in describing what specimens of a particular mineral often look like. Recognizing various habits helps a mineralogist to identify a remarkably large number of the known minerals.

Barrel—vanadinite from the Touissit mine, Morocco.

Prismatic—beryl from Pakistan.

Acicular—natrolite from Poona, India.

Plumose—barite from Cave-in-Rock, Illinois.

ded—celestine from Michigan.

Stellate—pyrophyllite from Mariposa County, California.

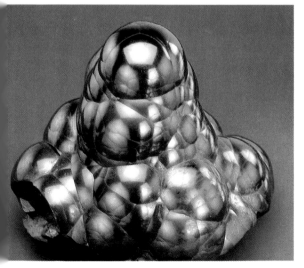

niform—hematite from Cumbria, England.

Micaceous—muscovite from Stoneham, Maine.

Oddities of Crystallization

Mineral crystals, breathtakingly beautiful in superb natural groupings, can also be mystifying, particularly when the forms they present are not readily understood or are not what they seem.

We know of many crystal aberrations because of the tendency of quartz to produce such an abundance of remarkable crystals. Most of these phenomena are surely present in crystals of other minerals, but other minerals are rarely as transparent and few ever reach the size of the quartz crystals that reveal unusual features.

Geodes

These crystal-filled rocks are round, but every round rock is not a potential geode. Geodes occur only in particular places, typically in rock units that extend over a broad area. Consequently, it is rare to find only one or two geodes: hundreds or even thousands in one site are more likely. Geodes usually occur in volcanic rocks, especially basalt, or in sedimentary rocks rich in carbonates of either calcium or calcium and magnesium, like limestone, dolomite, and some shales.

The amethyst geodes of Brazil and Uruguay are especially remarkable. This region is notable for the size of the producing area, which covers hundreds of square miles; for the abundance of geodes that have been and promise to be recovered; and for the size of many of the individual geodes. Hundreds are more than 2 feet in diameter or length, and some are even taller than a person. The crystals within the geodes, too, are generally larger than normal. Virtually all the geodes from this region contain amethyst-colored quartz.

In the past, miners attacked the geodes with sledgehammers. Specimen halves were almost unheard of until miners began to drill 1-inch "peepholes" to assess the quality of the amethyst inside. If the crystals are large and of good color, the geode is sledged into pieces and the amethyst sent off for faceting into gemstones. Smaller, paler amethyst crystals survive as mineral specimens. In this case, the miners chisel around the middle of the geode where they want it to split into halves, and eventually, with luck, it breaks in two. The results of the development of these skills are evident at any major mineral show where many perfectly opened, large geodes are on display.

Round geodes from Chihuahua, Mexico, are nicknamed "coconuts" and are usually only 5 to 7 inches in diameter. Traditionally, they were sawn in half with a diamond saw or split with a special cutting chain wrapped around the geode. Both techniques were very successful, but the locality, which produced thousands of geodes years ago, is now almost depleted. The quartz crystals in these Mexican geodes are varied: some colorless, others smoky, and a small percentage amethyst.

Geodes from sedimentary rocks are usually scarcer, even within the formations where they are known to appear. Generally, the collector must move a lot of rock before finding a geode, so the production has never rivaled that of the volcanic-rock sources. These geodes vary from fist-sized up to as large as one foot in diameter. The crystals are mostly quartz and they tend to be white, milky, or colorless.

Broken open like a coconut, the geode's crystalline interior is revealed. Light shining through this specimen from Warren County, Tennessee, imparts rich color to the white quartz.

Twins

It is not unusual for two or more crystals of the same mineral to be attached at some point. Sometimes, however, the angles between the individual crystals are not random. When crystals grow attached to one another at angles determined by their structure, they are known as twins. As is the case with geodes, only a limited number of localities produce twins, but those that do are prolific.

There are numerous types of twins in the mineral world. Some of these are associated with just one mineral or a few minerals. A common type often seen in quartz is known as the Japan Law twin because it was first observed in Japan, although it occurs in Korea and Arizona's Santa Cruz county. It is easy to recognize because the twinned crystals are almost always obviously flattened and meet at precisely 84 degrees 33 minutes in the plane of flattening.

These quartz crystals, from Kai, Japan, are twinned according to the Japan Law, which is a precise definition of the way the crystals meet.

Scepters

Normally, growth along the length of a crystal is uniform. If an elongated crystal suddenly or gradually fattens at one end, however, it becomes a scep-

Nature sculpted this marvelous grouping of amethyst quartz crystal scepters from Fulton County, Georgia.

ODDITIES OF CRYSTALLIZATION

The strange and fascinating forms, like the one above, taken by gypsum crystals from Naica, in Chihuahua, Mexico, are among the many oddities of crystallization.

ter. For reasons that are not fully understood—perhaps the solution in which the crystal was growing suddenly became enriched in silica, or the temperature or pressure changed—growth at or near the crystal's tip was favored over its stem. Scepters are most often seen in quartz, but they do sometimes occur in other minerals.

Phantoms

Like scepters, phantoms occur most often in quartz and are also produced by changes during crystal growth. Something, perhaps an earthquake or a volcanic eruption, disrupts the solution in which the crystal is growing, stirring it up and making it cloudy or muddy. Fine material dispersed in the solution then settles on the crystal and leaves a thin film over it. When the crystal resumes growing, it permanently incorporates the film.

The process may be repeated many times, resulting in a succession of films deposited at various stages of the crystal's growth. In transparent crystals, the thin coatings are visible as phantoms—smaller, ghostlike versions of the crystal's outermost faces.

Zoning

To observe zoning, a crystal usually must be cut into very thin slices for viewing under a microscope. This reveals a succession of concentric bands or zones, often distinguished by an alternation of

In transparent crystals, phantoms appear as smaller ghostlike versions of the outermost faces. Both these specimens are from Brazil. The sphere below was carved from a large crystal similar to the one on the left.

two colors or a progression through several colors, a different one in each zone.

Zoning is usually caused by frequent, often cyclic, changes in the environments where crystals grow. Some condition, whether chemistry, temperature, pressure, or acidity of solution, undergoes sharp change, so that areas of the crystal alter in composition in response.

Crystals of many minerals display complex zoning, and the effect can be quite beautiful.

Other Oddities

A number of the many oddities in the mineral world defy logic and have eluded explanation. One example is the strange, snakelike forms taken by gypsum from Naica, in Chihuahua, Mexico. These fascinating crystals twist and turn and then splay at the ends as though making a final scornful gesture in celebration of our failure to understand how or why this occurs.

Other exciting curiosities include quartz crystals pierced by pin-sized holes, sword-blade-shaped gypsum crystals with elongated cavities partially filled with a liquid that creates a natural carpenter's level, and diamond crystals with tiny red garnets trapped inside them. Under the microscope, tiny crystals of cuprite show mind-boggling right turns, while others mimic the skyscrapers of New York City. Minerals are, indeed, full of surprises and delights.

ZONING

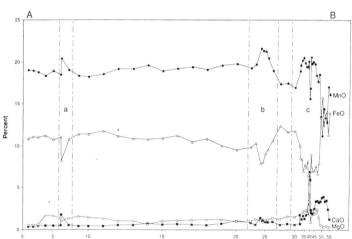

Shown above is a photograph of a thin section of eosphorite from a pegmatite in Brazil. Eosphorite is a manganese and iron phosphate mineral in which manganese is the dominant metal. The crystal is practically unzoned in the portion that is blue-green, but toward the later stages of growth it changes to red, orange, and yellow. Here its makeup fluctuates wildly, as the graph of a chemical analysis of the crystal from one end to the other demonstrates.

At the last period of growth, the fluctuations became quite violent. In two very thin zones the mineral contains more iron than manganese. The crystal also contains modest amounts of calcium and magnesium, remarkably constant components throughout, again until the last growth period, during which their content increases abruptly.

Having neither observed this growth nor seen the specimen before it was removed from its matrix, we can only guess at the cause of this extreme zoning. It is likely that it was brought about by some physical disruption of the opening in which the crystal was growing. Scientists who study pegmatites have concluded that heat emitted during crystal growth increases pressure within such openings, causing the "pockets" to rupture and actually explode. When this occurs, new fluids with different compositions can flood in, and the pressure drops, too. Either change could have contributed to the variations seen in this crystal.

Replacements

The expression "turned to stone" is not pure fiction in the mineral world, where it is synonymous with some processes of fossilization. After a plant or animal is buried, minerals may replace all of the organic tissues, or only such hard parts as shell and bone. Sometimes the substitution is not obvious. With some fossil mollusks, for example, another form of calcium carbonate replaces the original calcium carbonate shell secreted by the animal.

When woody plants die and fall to the ground, they usually decay and effectively disappear, but occasionally they may be buried under mud, sand, silt, or volcanic ash while still fresh, perhaps by a cataclysmic event. The violent eruption of Mount St. Helens in Washington, for instance, blew down a large number of trees and immediately engulfed them in a tidal wave of mud and ash.

Even under such conditions of burial, most trees will decay, leaving no trace. If favorable conditions prevail, however, decay is somewhat circumvented; instead, over a long period of time a mineral (usually some form of silica, either quartz or opal) replaces the wood. The cell structure may be preserved in detail so perfect that it is just as apparent as in a fresh cut of a live tree.

Mineral replacement in wood takes place in several steps. The walls of the cell are hard cellulose and lignin, while the pulp within the cells is soft and porous. Pulp decays and disappears before the cell walls begin to decompose. Under ideal conditions, mineral first replaces the pulp, then the cell walls. As a result, this petrified wood preserves original cell structure in sharp definition, making it possible to identify the living tree from its fossil. As with a live tree, a petrified log must be sliced like a loaf of bread to obtain a cross section that reveals the cell structure. For display purposes, these cross sections are highly polished.

In many examples of fossil wood, patterns of cell structure are not sufficiently distinct for the tree to be identified. This is unfortunately true of most of the fossil logs in the most famous find of petrified wood, Petrified Forest National Park in northeastern Arizona. The logs are still remarkable, how-

The beautiful specimen above is oak wood that has been replaced by common opal. It was found at Clover Creek, Lincoln County, Idaho.

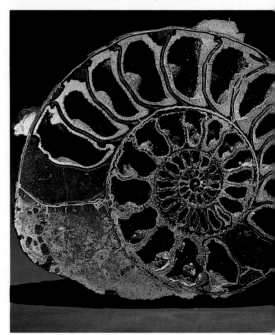

The calcium carbonate of the shell of this ammonite, an extinct relative of the chambered nautilus, was replaced by pyrite, an iron sulfide. The fossil was found in Lukow, Lublin, Poland.

ever, because they are beautifully mottled in patches and splotches of pink, lavender, yellow, red-blue, and brown to black (although the coloring has nothing to do with the original wood). It is possible to tell from the logs' shapes that most were araucarian, a soft wood; perhaps that is why cell preservation is poor. Many other types of fossil plants are also found in the park; more than thirty species have been identified in this vast area.

Shellfish are also candidates for replacement. When buried in sediment—especially mud that is rich in iron and possibly sulfur—iron sulfide in the form of pyrite or marcasite may replace the shells. Ammonites, extinct relatives of the chambered nautilus, were sometimes preserved when pyrite somehow permeated their shells. The external shell may not appear to have changed at all, yet iron sulfide has penetrated the chambers so that pyrite precipitated in crystalline layers within each of them. When the fossil is cut in half and polished, the contrast of the bronzy metallic pyrite coating the chamber walls of the ammonite's gray shell can be striking.

Most of the more common fossils are preserved through replacement by one of a small number of minerals: calcite, aragonite, quartz, pyrite, marcasite, apatite, goethite, and hematite. Graphite should probably be included, too, because many plants and soft body parts are carbonized. Not all fossils are preserved through replacement. Some fossils are simply the imprint of an organism's body in mud or silt that later solidified into rock; all other traces of the creature have disappeared. In addition, insects, and even frogs, have been preserved in amber, and large animals have been preserved in, but not replaced by, asphalt, as can be seen at the La Brea tar pits in Los Angeles County, California.

Novice collectors may be fooled by specimens of pyrite and other minerals that mimic fossils but actually have no relationship to an organism. Since crystals can grow into bizarre and fantastic forms, occasionally these forms accidentally resemble creatures of the plant or animal world. Who knows why pyrite sometimes looks like a mushroom? Probably no one. Pyrite that looks like sand dollars and dendritic, or treelike, manganese oxide minerals are more easily explained. In each case, the mineral grew within flat, narrow openings in thinly layered sedimentary rocks. Confined to a single plane, it spread out as far as the space available allowed, growing larger as long as its supply of mineral "nutrients" lasted.

"SAND DOLLARS" AND DENDRITES

A pyrite "sand dollar" from Sparta, Illinois.

Pyrite sand dollars are common in certain carbonaceous shales (sedimentary rocks produced by compacted silt rich in carbonized plant material). In these mineral mimics, growth began at one point and proceeded at a constant rate in all directions in a radial pattern, with the individual crystals appearing as spokes in a bicycle wheel.

Dendrites—treelike growths typically of manganese oxide minerals—also began growing from one point, probably the juncture of a bedding plane and a cross fracture. As the crystals grew, they fanned out as best they could in the narrowest of openings in sandstone, a sedimentary rock composed almost entirely of quartz that was once sand. A force much like capillary attraction must have drawn the branches out until there were no more dissolved nutrients to permit the crystals to grow.

Dendritic crystals of the mineral psilomelane, from Solenhofen, Bavaria, Germany.

Physical Properties

A mineral can be identified by a number of physical properties, including hardness, luster, color, cleavage, fracture, and specific gravity. Identifying the more than three thousand mineral species known to science ranges from simple to exceedingly difficult.

Beautifully developed crystals of the most common and familiar species—there are at least several hundred in this group—do not present much of a challenge. Mineralogists and experienced collectors recognize them on sight or after applying relatively simple tests. But identifying poor crystals of even the common minerals, or samples of less common minerals, including good crystals, may require great knowledge and skill, as well as highly sophisticated technology. Chemical analysis and X-ray diffraction are usually available only at larger museums and universities. Even these institutions can ill afford to use these techniques for every identification because of the time and the cost involved. Therefore, in many cases, it is necessary to rely on other tests designed to give specific information about a mineral's physical properties. A collector, even with minimal equipment, can apply several of the tests that observation of these properties requires and thereby greatly reduce the number of candidates. It is far more satisfying to take a short list of possible identifications to an expert than to hand over an unknown substance and say, "I haven't the least idea of what this could be."

Hardness

One of the first physical properties of a mineral to be assessed is its hardness. The most commonly used standard is the Mohs scale of relative hardness.

Although it resembles a piece of modern sculpture, the specimen, left, from Ambassaguas, Spain, is nature's handiwork. The perfectly regular arrangements of iron and sulfur atoms that make up pyrite dictate the crystal's cubic form.

Diamond, the hardest of minerals, is number 10 on the Mohs scale. This 230-carat diamond crystal, the Oppenheimer, is from the Dutoitspan mine near Kimberley, in South Africa.

Frederich Mohs, a German mineralogist, devised this scale in 1824. Using diamond, the hardest of minerals, and talc, one of the softest, as extremes, he then selected eight more common minerals and numbered the ten in order of increasing hardness. Talc, of course, is number 1, followed by gypsum, calcite, fluorite, apatite, orthoclase, quartz, topaz, corundum, and diamond.

Relative hardness is tested by trying to scratch one mineral with another, since any mineral will scratch all other minerals that have a lower hardness number. Fluorite, for example, will scratch calcite and gypsum, which are softer, but will not scratch apatite, which is harder. An unidentified mineral can be tested for hardness by applying a series of scratch tests until it can be determined where on the Mohs scale it belongs.

While the differences in hardness represented by successive numbers on the Mohs scale are almost equal from numbers 1 to 9, diamond is many times harder than corundum. The difference in fact is far greater than between corundum (number 9) and talc (number 1). A more accurate comparison of hardness can be made by using an indentation test. In the Knoop test, for example, dents made in a mineral by a diamond point at controlled pressures are carefully measured. Such tests do, however, require complex and expensive instruments.

Luster

Terms describing luster indicate the way a mineral's surface interacts with light. Minerals that look like metals, for example, are described as having a metallic luster. Slightly less metallic-looking minerals are called submetallic. Most metallic and submetallic minerals are opaque, meaning no light can be transmitted through them.

The remaining minerals are considered nonmetallic. Of these, the most intense reflectors of light are said to be adamantine, meaning diamondlike. Less brilliant reflectors are vitreous, meaning glassy. Poor reflectors with dull lusters are described by a variety of terms—earthy, waxy, silky, pearly. Although all of these terms are quite subjective, their use helps to characterize a mineral and observations of luster are worth noting.

Stibnite, an ore of antimony, has a metallic luster. This group of crystals is from Iyo, Japan.

PHYSICAL PROPERTIES

Mimetite is found in a variety of colors. This yellow specimen is from San Pedro Corralitos, in Chihuahua, Mexico.

Color

The most obvious of a mineral's properties is its color. A mineral may have an inherent color, dictated by its essential chemical makeup, or it may have an imparted color. Not seen in pure samples of the mineral, imparted color is introduced when a minor amount of another element is present.

Most copper minerals, for example, are usually intense blue or green. These are normal, inherent colors for copper minerals, at least those that also contain oxygen. With sulfur, copper ranges from golden to silvery or black. Sulfides (compounds with sulfur) and related species of most metals are opaque, metallic to submetallic in luster, and not particularly colorful.

Oxides (compounds with oxygen) of various elements combined with metals tend to be more colorful. In these compounds, cobalt may result in a

Azurite, a bright blue mineral, and malachite, a rich green mineral, are both copper compounds and are often found together. This specimen is from Bisbee, Arizona.

Manganese gives rhodochrosite its red or pink color. This spectacular specimen comes from the N'Chwaning mine, Cape Province, South Africa.

pink mineral; iron, in yellow; nickel, in green or pink; and chromium, in green. This is the case when the metal is either an essential constituent or merely a contaminant. The primary difference is the intensity of the color. Many minerals would be colorless if they were pure, but contaminants readily impart tints of color. Calcite, composed of calcium carbonate, can easily accommodate tiny amounts of other metals in its structure. It responds to the metal it adopts by being pink, blue, green, or yellow; whatever the appropriate color for that contaminant may be.

Other Optical Properties

Luster and color are only two of a mineral's optical properties. Detailed observation of many other optical properties requires special equipment, preferably a microscope. Not surprisingly, like many of a mineral's other properties, its optical properties are related to or affected by its structure.

Highly symmetrical minerals, those with a structure based on the cube, are called isotropic, which means that light passes through them in the same way in all directions. Other minerals are anisotropic, meaning that their optical characteristics are not the same in all directions. This is well illustrated in the gem variety of spodumene, known as kunzite, which absorbs light differently in each of three orientations, making it appear three different colors. It is light lilac from one direction, intense magenta from another, and green from the third. Still another excellent example of an anisotropic mineral is the gem variety of zoisite, which is called tanzanite. It is sapphire blue from one direction, magenta from the second, and deep red from the third.

Zoisite, from Tanzania, is an anisotropic mineral—absorbing light differently in each of three orientations, thus appearing a different color in each. Two of these are shown above.

Cleavage

Different minerals may come apart, or cleave, in different ways. This fascinating property is dramatically demonstrated by the mica group of minerals. The most common of the micas is muscovite, a mineral that was widely used for insulation in toasters and as windows for high-temperature ovens before synthetic materials replaced it. Muscovite is well suited for these uses because it easily splits into paper-thin, flexible sheets. Micas cleave because the crystal's structure contains alternate layers of atoms with bonds so weak that they can easily be pulled apart, but in only one direction. The bonds in all other directions are strong. Consequently, thin sheets of mica are extremely tough.

Some isometric minerals also cleave, but as their name implies, in these minerals there are several identical cleavage directions corresponding to the symmetry of the crystal. Galena (lead sulfide) has cubic cleavage, which means that it breaks along smooth planes parallel to the faces of a cube. Fluorite (calcium fluoride) has octahedral cleavage. It breaks perfectly in four directions, and the planes are parallel to those of an octahedron.

Calcite is another mineral with perfect cleavage. Like galena, it cleaves in three directions, but the angles between the planes are not 90 degrees. Cleavage produces a rhomb, a solid whose faces are all equilateral parallelograms. Calcite is brittle and will shatter if struck with a hammer, but all the pieces will break along the same cleavage directions and all will be rhomboid. Many other minerals possess less perfect cleavage, whether in just one direction or in two or even more directions.

Fracture

Fracture describes how a mineral breaks other than along natural cleavage planes, in terms that refer to the surface of a newly broken crystal. The surface may be very smooth and curved (conchoidal), rough and irregular (hackly), splintery, even, or uneven. Many minerals display both cleavage and characteristic fracture.

Parting

Another term used in descriptive mineralogy is parting, which refers to the manner in which some minerals break along smooth planes that are not cleavage planes. A mineral can part along planes where changes in chemical composition have made weaker atomic bonds than elsewhere in the crystal. Unlike cleavage, which can occur anywhere parallel to a cleavage direction in a crystal, parting occurs only where layers of different chemical composition exist.

Specific Gravity

A cubic centimeter of a heavy mineral like topaz, galena, or barite weighs far more than a cubic centimeter of a light mineral like gypsum or sulfur. The relative "heaviness" of a mineral is measured as the specific gravity, which relates the mineral's weight to that of an equal volume of water.

Specific gravity can be determined quite precisely, but specialized equipment is required, and although it is not prohibitively expensive, it is beyond the reach of most collectors.

Other Properties

How a mineral breaks (cleavage or fracture), its relative hardness (resistance to scratching), and its specific gravity (density relative to water) all repre-

Calcite perfectly illustrates cleavage. When a crystal is broken, as shown above, all the pieces will be rhombohedral.

sent generally observable physical properties. Other properties including luminescence, magnetism, radioactivity, and tenacity are generally less useful for identifications except in special cases.

Luminescence is the way a mineral responds to ultraviolet light. Some minerals emit light of beautiful and intense colors when exposed to ultraviolet light. If this effect stops when the light source is removed, it is called fluorescence. If a mineral continues to emit light after the ultraviolet light is removed, it is called phosphorescence.

Only a handful of minerals are magnetic. The intensity of the attraction may, however, be useful in distinguishing between them.

Radioactive minerals contain unstable isotopes of uranium or thorium and so emit radiation that can be detected with a Geiger counter. Radioactive minerals will also expose photographic film, providing a simple and inexpensive test. There are, perhaps, two hundred radioactive minerals. Some, like uraninite and thorianite, are always radioactive because their chemical compositions inherently contain uranium or thorium. Others are merely contaminated with small amounts of radioactive elements not essential to their basic chemistry.

Tenacity describes how a mineral responds to mechanically induced changes of shape or form. Minerals that bend easily are said to be flexible. Those that bend and return to their original shapes are elastic. Sectile minerals can be cut with a knife without producing tiny fragments or powder. It can be quite useful to compare qualities like brittleness and malleability. Gold, for example, may be readily distinguished from nearly all other gold-colored metallic compounds because it is malleable and they are brittle. If a piece of gold is hammered, it will flatten, while the others will break into tiny pieces or even powder—a simple but effective test.

This specimen of zinc ore, from Franklin, New Jersey, contains willemite and calcite, minerals which exhibit fluorescence. When exposed to ultraviolet light, above, the willemite emits a vibrant green light and the calcite an intense red light.

Classifying Minerals

Chemical compounds do not lend themselves to the kind of hierarchical classification systems that have been adopted for plants and animals. Minerals cannot be broken down logically into phylum, order, family, genus, and species. Despite obvious difficulties, some early mineralogists nevertheless did attempt unsuccessfully to force such a system on minerals.

In 1816, however, a Swedish chemist, J. J. Berzelius, introduced a classification scheme based on the chemistry of minerals. By the middle of the nineteenth century the American mineralogist James Dwight Dana had refined Berzelius's system, and Dana's system of mineralogy has been widely used ever since. Today, virtually all the major mineral collections of the world, including that of the National Museum of Natural History, are arranged according to the Dana system. The basis of the system is the division of minerals into classes according to similarities in chemical composition.

The first class, native elements, includes minerals found as individual elements uncombined with other elements. As minerals most of them are called by the name of the same element: copper, lead, silver, gold, mercury, arsenic, and sulfur, for example. Others, such as diamond and graphite (both varieties of carbon), have distinctive names. As is the case with all minerals, native elements are not expected to be free of impurities; they are all "contaminated" to a lesser or greater degree with other elements.

Sulfides, a class of simple compounds, are metals combined with sulfur, arsenic, antimony, bismuth, tellurium, or selenium. Pyrite, for example, is iron and sulfur; galena is lead and sulfur; cinnabar is mercury and sulfur. Some nonsulfur "sulfides" include nickeline (nickel arsenide) and krennerite (gold telluride). Most of these minerals have a metallic luster.

Minerals of the third class, sulfosalts, are similar to sulfides, especially in physical properties (most members of this class also have metallic luster).

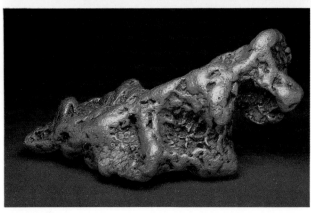

These specimens are examples of native elements: sulfur, left top, from Italy; gold, left, from California; and copper, above, from Michigan.

Less familiar than sulfides, sulfosalts are composed of the same elements, but in somewhat more complex chemical combinations. Few sulfosalt minerals are known by name to noncollectors.

Oxides, the fourth class of minerals, are relatively simple compounds of metals with oxygen. One exception is ice, which is hydrogen (a nonmetal) and oxygen. Perhaps the best known of the other oxides is corundum, an aluminum and oxygen compound, which in its gem forms is called sapphire and ruby. Because of corundum's great hardness, exceeded only by diamond, it is also an important abrasive. Mixed with garnet it is known as emery.

The next class is the hydroxides, oxides with water. Then there are multiple oxides, which are another step up in complexity.

The halides are metals plus one or more of the halogen elements—chlorine, fluorine, bromine, or iodine. Two familiar minerals in this class are halite (sodium chloride), the same compound as table salt, and fluorite (calcium fluoride), an important flux used in smelting to lower the melting points of metals.

There are a number of other classes, each named after a complex of an element and oxygen in combination with one or more metals. These complexes include the carbonates, phosphates, nitrates, borates, arsenates, and sulfates.

Finally, there are the silicates, by far the largest and most diverse class. Silicon and oxygen form a structural unit that can bond to other elements in an extraordinary variety of ways, forming a large number of distinct structural types, each a different mineral. Most museums tend to place quartz (silicon dioxide) in the silicate class, but a reasonable argument can be made for classifying quartz with the simple oxides.

There are, of course, deficiencies in any classification system. After all, minerals, unlike living organisms, did not evolve from simple to complex, with genetic links between them and common ancestors. Imperfect though the Dana system may be, it does achieve the most important objectives: to bring a degree of order to the objects of the mineral kingdom so more sense can be made of it, and to be able to file these objects away while assuring quick and easy retrieval. Dana's system makes both possible about as efficiently as can be expected, and that is why it is the system used by most museums.

Halite, above top, and fluorite, above, are halides—metals plus halogen elements. (The halite specimen is from Rhodes Marsh, Nevada; the fluorite from Coahuila, Mexico.)

Quartz belongs to the silicate class of minerals. This amethyst quartz is from Iron Station, North Carolina.

Mineral Groups

Minerals themselves defy effective overall classification, but mineralogists recognize certain groups, or families, whose members are sufficiently similar to be conveniently lumped together. The unifying links are usually related to structure and chemical composition. Limited groupings may also be based upon close physical similarities. Since identical structures may be found in different classes, and since minerals of the same composition may differ in structure, it is impossible to devise a system into which all minerals fit perfectly. Compromises can become a necessity in grouping minerals.

An interesting example is a group based on the structure of a mineral no longer even considered a distinct species—apatite. Years ago, apatite was defined as a calcium phosphate mineral having a particular hexagonal structure. It was also known to contain fluorine, chlorine, or hydroxyl or some combination of them. Today, any apatite with more fluorine than chlorine or hydroxyl is called fluorapatite. Its analogues are either chlorapatite or hydroxylapatite. These three minerals constitute a neat little family, but the group also takes in other minerals, some of which are not calcium phosphates, but all of which have the same apatite structure. They include pyromorphite (lead phosphate), vanadinite (lead vanadate), and mimetite (lead arsenate), all of which also contain chlorine. Al-

though these three are placed in the apatite group, vanadinite and mimetite (unlike other group members) do not belong to the phosphate class.

There are many other important mineral groups. Spinels and garnets are two of the best-known and the simplest, meaning that the basic chemical formula for the family is not complex. The spinels, for example, all have formulas that can be represented as AB_2O_4, where A and B are metals. The A elements commonly are cobalt, copper, ferrous iron, magnesium, manganese, tin, or zinc. The B elements typically are aluminum, chromium, ferric iron, or vanadium. In this case, the group name, spinel, is also that of a valid member of the group with a formula $MgAl_2O_4$.

The garnets are not too different from the spinels except that they are silicates with the general formula $A_3B_2(SiO_4)_3$. Many of the same metals found in the A and B positions in spinels occupy the A and B positions in garnets. The name garnet applies only to the group, which consists of fourteen members; there is no garnet mineral species. The most common mineral of the garnet group, almandine, has ferrous iron in the A site and aluminum in the B.

Within the silicate class there are several very large and extremely complex groups, especially the amphiboles, micas, pyroxenes, and zeolites. The members of most of these groups share certain structural traits, but because the basic chemical formula for each group contains many components, there are numerous possible permutations in chemical composition.

The zeolites are somewhat unusual. Members of this group do not all have similar structures, but they are all silicates that share the ability to gain or lose water or to exchange water for ions. This makes zeolites excellent water "softeners" because they can selectively extract undesirable ions from contaminated water.

These three minerals belong to the apatite group: left, fluorapatite, which contains more fluorine than chlorine or hydroxyl; above, pyromorphite, lead phosphate; right, vanadinite, lead vanadate. (The fluorapatite specimen is from Panasqueira, Portugal, the pyromorphite from Les Farges, France, and the vanadinite from Mibladen, Morocco.)

Modern Mineralogy

The science of mineralogy has progressed dramatically since the 1960s because of technological advances. Possibly the most profound change lies in the approach to research. Earlier studies were directed toward gaining a better understanding of minerals themselves, principally of their composition and structure. Modern researchers try to discover what a crystal can reveal about its early history.

Minerals are amazing reservoirs of information that is locked away until an enterprising scientist develops a means of coaxing it from them. Some minerals are time clocks, making it possible to date important events in their pasts, including their primary crystallization. Other minerals provide information about the pressure under which they grew; this can be translated into a measure of the depth below the earth's surface at which they formed. Still others carry clues to their temperatures of formation. Some may carry remnant polarities, which record the orientation and intensity of the earth's magnetic field when their crystallization occurred. Minerals may give evidence of violent episodes in their past, such as meteorite impact. Zoning within individual crystals permits interpretation of less dramatic changes that may have occurred in the environments of crystallization, perhaps a gradual lowering of the temperature range through which a crystal grew.

Mineralogy has progressed from simple measurements of such obvious properties as hardness, crystal shape, light transmission characteristics (or optics), and density, to esoteric studies of isotopic composition, crystal structure analysis, variations in thermal and elastic properties with direction and under changing conditions, and infrared absorption properties.

The equipment that mineralogists use has similarly evolved from unsophisticated balances for determining density, goniometers for measuring crystal angles, and classical laboratory paraphernalia for standard chemical analyses. Modern mineralogists have at their disposal a wealth of elaborate, expensive, and complex instruments capable of doing things not even dreamed of twenty or thirty years ago. Two of the most important and well established are the electron microprobe and the single crystal X-ray diffractometer, both of which have benefited from a number of improvements since their introduction. Today, in addition, the analytical scanning and transmission electron microscopes, infrared spectrometers, atomic absorption and emission spectrometers, and a variety of X-ray diffraction instruments, are wedded to computers for rapid analyses of the data they gather.

It is interesting to look at the way one of these instruments, the electron microprobe, is used in mineralogical studies. The probe, as it is familiarly known, utilizes a high-energy electron beam that is focused to a diameter of about 5 microns (1 micron equals 1 millionth of a meter or 0.000039 inch). The beam strikes a mineral sample with so much energy that the elements within it are forced to emit X-rays. The X-rays of each element have their own characteristic wavelength, and the intensity of the wavelength is directly proportional to the amount of the element present. To learn how much of each element is present, a mineralogist need only collect and measure the intensity of the X-rays being emitted. The technique is fast and largely nondestructive, so the sample usually survives the analysis.

Since such a tiny area is analyzed, the sample can be moved around under the beam and different parts can be analyzed independently. Formerly, with so-called wet-chemical analysis, large samples were required, and each was run in bulk so that the

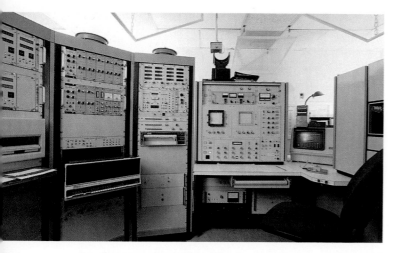

An electron microprobe, like the museum's, permits rapid nondestructive quantitative chemical analyses of tiny crystals.

This is a scanning electron microscope image of cuprite crystals from the Magma mine in Superior, Arizona. The tiny crystals are not visible except under extreme magnification.

entire sample was analyzed. Because the composition of most crystals is not uniform, what was obtained was an average composition for the entire sample. The probe, however, makes it possible to analyze the many different zones within a microscopic crystal so that changes in composition can be observed and diagrammed. In addition, by expanding the electron beam, the X-ray emission may be photographed element by element, producing a succession of images that clearly define the zones within a single crystal or chart the configuration of the members of a mixture of two or more minerals.

Despite fantastic advances in technology and the marvelous types of studies possible today, the less exciting and now more or less routine description of new mineral species is still actively pursued. The number of species is, therefore, continually increasing. The method of introducing a new species is neither difficult nor mysterious.

After finding that a natural substance may be a new mineral, the next step is to gather certain basic information about it. Its chemical composition must be determined, and its X-ray diffraction data recorded. With these two parameters, it is usually possible to tell whether there is already a known mineral of its type. If it appears that there is not, then the measurement of as many of the substance's other properties as possible is undertaken.

When all available data have been derived from the sample, an abstract of the data is prepared and submitted for consideration to an international commission of mineralogists. During the review, any member of the commission may request additional data or challenge any of the data. At some point, the commission votes on the proposal. If a satisfactory majority approves, the mineral is accepted. The full description may then be submitted to a journal for publication.

It is customary to select a name for the mineral before submitting the abstract, so the name can be considered at the same time as the description, although in a separate vote. Sometimes the description may pass, but the name choice may fail. Usually this is because the proposed name is too similar to an existing name, with the potential for confusion. The proposer must then select another more suitable name.

Collecting Minerals

Any sample of a mineral that has been acquired for study or placement in a collection is referred to as a "specimen." The term does not necessarily imply field-collected, although specimens included in reference or study collections were often collected in the field rather than purchased from a dealer. Because they were not selected for exhibition, they tend to be less exquisite than specimens displayed in museums. Field collecting is almost entirely done by professional earth scientists and their students.

Other collectors specialize in aesthetic specimens, either beautiful crystals or materials that could be or have been altered in form and polished. Most of these specimens are acquired by purchase from dealers or through exchange with other collectors or museums. The value of display specimens has greatly escalated in recent years, and the best mineral specimens, such as those shown in museums, now command extremely high prices.

When considering exhibition-quality specimens, great emphasis is placed on overall attractiveness. Handsome groupings of crystals are especially admired. Specimens with sculptural qualities are eye-catching. Color, too, is an important consideration. Specimens with lovely soft or vibrant colors sell themselves.

An otherwise valuable specimen can lose most of its value if any of the more prominent crystals has been scratched or damaged, just as a fine piece of china or an important oil painting drops in value when it has been abused. Consequently, fine specimens should be handled with extreme care.

Most exciting, valuable mineral specimens have been accidental by-products of mining, usually for something other than those specimens. Many of the world's great malachite and azurite specimens, for example, have come out of the immense underground copper mine at Bisbee, Arizona. Another large and important mine at Tsumeb, Namibia, may have produced the greatest number of exhibition-quality specimens of any locality. Close to one hundred minerals, in thousands of beautiful specimens, have been found deep underground in this mine, which has been worked almost continuously since it produced its first ore in 1900.

Recovering these wonderful specimens begins when the tunnels and drifts of a mine encounter crystal-lined openings in the rock. Because of the crystals' fragility and the violence of mining, only a small number of the crystals may survive. It is heartbreaking for a collector to have to walk upon loose crystals that have been dislodged from the walls and ceilings in the course of mining or by natural forces, but it is often necessary in order to get to intact crystals that remain to be collected. Not only must they be carefully removed from the surrounding rock, they must also be protected by skillful packing and then carried out of the mine, sometimes even in a miner's lunchpail, a procedure widely romanticized but probably not often used.

A few mines, particularly those that produce gem materials, operate specifically to obtain specimens. Gem mines are worked primarily for cutting material, but they also often produce, as a by-product, crystal specimens more valuable in their natural form than they would be as gemstones. The return on gemstones is, however, usually much greater than on crystal specimens. Mining for display specimens of crystal groups is only rarely profitable.

It is axiomatic that the best specimens are found by the people who constantly work a deposit. Because excellent specimens only appear sporadically, a collector cannot hope to visit a locality for just

Mineral collectors at the Foote Mineral Company's spodumene mine in Kings Mountain, North Carolina.

A geologist collects specimens of metamorphosed volcanic rock for radiometric age dating at the Mt. Goddard pendant of the Sierra Nevada batholith in California.

part of one day and find a "pocket" of superb specimens. This is why it makes more sense for a museum curator to buy a fine specimen from a dealer than to spend as much or more money on a usually futile trip to the place where the specimens are known to occur.

Amassing a collection is only part of what mineral collecting is all about. A responsible collector carefully catalogs each specimen, recording all pertinent details such as specifics of the locality where it was found and its history (including who collected the specimen, when, and former owners), as well as the source, cost, and date of its acquisition. Specimens must always be labeled; to avoid getting labels confused, each specimen should have a catalog number physically applied to it, the same number that appears on the label. Just as damage to a specimen can drastically diminish its value, a poorly cataloged and labeled collection can lose much, if not most, of its value. Mineral specimens are a limited commodity, and their preservation should be a matter of concern to all collectors.

45

The National Mineral Collection

The United States National Museum of Natural History founded its mineral collection in 1859, although its parent institution, the Smithsonian, had a collection more than a decade earlier. When James Smithson, a wealthy English chemist, bequeathed £106,000 to the United States for the creation of the Smithsonian Institution, his legacy also included his large mineral collection. According to an account of the time:

> Among the effects of the late Mr. Smithson is a cabinet which, so far as it has been examined, proves to consist of a choice and beautiful collection of minerals, comprising probably eight or ten thousand specimens. The specimens, though generally small, are extremely perfect, and constitute a very complete geological and mineralogical series, embracing the finest varieties of crystallization, rendered more valuable by accompanying figures and descriptions by Mr. Smithson, and in his own writing.

Tragically, in 1865 a fire on the second floor of the Smithsonian building (known as the "Castle") totally destroyed Smithson's collection and any catalog of it that may have existed. How exciting it would be if the collection had survived and could be viewed today; how much it would tell us about Smithson's collecting instincts.

The bronze Smithson medal was created by Paul A. Vincze of London.

The Collection Today

The National Museum collection began with only 793 specimens. By 1880 the collection had grown to 3,551 specimens, and over the next fifteen years it exploded to 45,291 specimens. It did not take its place among the great world collections until 1926, when the museum received two of the best private American collections—the Frederick A. Canfield and the Washington A. Roebling collections, both of which were accompanied by substantial endowments. Canfield and Roebling were contemporaries who lived about fifty miles apart in New Jersey; ironically, both died in the month of July in 1926.

The Canfield and Roebling collections together numbered about 25,000 specimens. Another fifty private or institutional collections have been incorporated into the national collection since 1926, about thirty-five of which have been acquired since 1950. Today, the Smithsonian's collection is regarded as one of the world's preeminent. Perhaps as few as six museums share this distinction. The Smithsonian collection is extraordinarily balanced because equal attention has been given to all aspects of its growth, at least during the last thirty years. Some of the credit may be attributed to federal law, which requires most other government agencies to turn over specimens to the museum, and which allows a tax deduction to donors of specimens or money. Without these inducements, the collection would certainly be much less important than it is today.

Estimating the exact size of the collection is difficult and not particularly meaningful. There are almost two hundred thousand catalog entries, but a single entry often represents a multitude of pieces. Today the total number of specimens far exceeds a quarter of a million.

Three Collections in One

The National Museum collection comprises three fundamentally different subsets of specimens: the reference, type, and exhibit collections. The largest is the reference collection. Among its hundreds of thousands of study specimens, several thousand are "described" specimens, which have been analyzed or tested and for which some form of data is main-

Smithsonite, a zinc carbonate mineral, was first recognized by James Smithson, a chemist and a collector of minerals. This specimen is from the Kelly mine in New Mexico.

tained on file. Usually this record is a chemical analysis, but it can also consist of descriptions of optical or physical properties or other information.

The museum is extremely active in providing samples for scientific research, sending many hundreds all over the world each year. When a scientist requests a sample, generally only a tiny fragment is needed. Of greater interest than the largest, most perfect of crystals are specimens that show variations in form or color and also preserve the relationship between the particular mineral and its associated minerals, the matrix. Since small chips usually suffice, the museum seldom gives up whole specimens—fortunately, since many studies performed result in the destruction of the sample.

The museum benefits from providing research samples, not only from acknowledgements in published papers, detailing the results of the studies, but particularly in gaining valuable information about the specimens from which the chips were taken. This information is carefully recorded, filed, and cross-referenced to the specimens. In this way, a substantial volume of detailed data relating to specimens in the National Museum collection is gradually amassed, making them "books" in the mineral library. Their value and potential usefulness to future investigators has been greatly enhanced.

Kept physically separate from the reference collection is the type collection of specimens used in describing species new to the science of mineralogy. The more than one thousand type specimens in this valuable subcollection represent nearly one-third of the total number of mineral species known. Of all specimens in the national collection, the types are perhaps the most important scientifically. If questions arise about the validity of a species, type material is the ideal standard used in confirming or challenging any detail of the mineral's description. Of course, type specimens have not always been preserved; in some cases, modern studies of older species must be performed on specimens assumed to be similar to the type, but the relevance of the results can always be challenged when type material is not used.

Behind the scenes in the museum, thousands of reference specimens are carefully stored in drawers.

Finally, there is the exhibit collection of minerals used for display and others that are particularly valuable because of extraordinary form or beauty. Educating and intriguing the public are unquestionably important goals of all museums, and it is with these objectives in mind that the National Museum of Natural History acquires such specimens, mostly through purchases with endowment funds. (No federal money is appropriated for purchasing specimens for either the mineral or the gem collection.) These specimens are so valuable that they are kept in a high-security storage facility unless they are displayed to the public.

Most of the redundancy of display-quality specimens has grown from continuing efforts to upgrade the collection. When superior additions supersede older specimens, the older ones, which are still very fine, must be stored securely. There would be little point in trying to display all the very good specimens of every mineral. Even if there were ample exhibit space, the volume of specimens would be so overwhelming that visitors might find it difficult, if not impossible, to appreciate the truly fine examples that the museum hopes to make memorable.

Only a fraction, at best, of the entire National Museum collection is on display. Most of the specimens carefully stored in drawers behind the scenes —about 90 percent of the collection—probably would not be considered exciting in an exhibit. Many reference specimens are blocky pieces, uninteresting in form. Others are merely tiny fragments of relatively rare minerals, each in its own small box. Nevertheless, these specimens are often the truly useful material of mineral science.

Building the Collection

Places where minerals are found, called localities, are almost always short-lived in terms of specimen production, so specimens must be acquired while they are available. Many mineral deposits tend to be small, and they are often totally mined out upon discovery, leaving nothing at the site for future collectors. Because the majority of mineral deposits lie near the earth's surface and are rather easily discovered, it may be assumed that most have already been found. There is not an infinite supply of specimen material in the earth. The conditions that produced a particular mineral deposit may be unique to one site; consequently, the minerals associated with it are unlikely to appear anywhere else.

Even if this were not so, few mineralogists can embark on collecting expeditions to find all the materials needed for scientific studies. Most researchers rely upon major museums for what they must have. If museums fail to meet this challenge, mineral samples will not be obtainable and the progress of mineralogical science will be seriously impeded, as well as all kinds of research beyond pure mineralogy. Samples from the National Museum collection have been provided to scientists for such wide-ranging interests as the search for new ion filters, use as pigments, and dental research.

Specimens for the reference collection are acquired in many ways. Unlike exhibition specimens, study specimens are seldom purchased. Because other government agencies, including the U.S. Geological Survey, are obliged to turn over the specimens they collect, many fine specimens have come to the collection in this way. In addition, many of the scientists who describe new minerals are aware of the importance of depositing type specimens in a major museum, and the National Museum is often the beneficiary. Researchers who obtain samples from the museum frequently return the favor by contributing mineralogical materials to the collection.

The museum's curators do collect some reference specimens in the field, but generally only from areas that have been the subject of continuing study by a particular scientist. The investigator attempts to find specimens that may reveal important information relating to the conditions under which the minerals at that site came into existence. Such studies often last for years and involve many visits to the locality.

Specimens coming into the national collection are cataloged at the earliest opportunity. Years ago entries were handwritten in ledgers. Today they are entered into a computer. A number is assigned to each specimen, then physically applied to it, and a label prepared to accompany the corresponding specimen. The final step is to distribute the specimens into the collection. When an entire collection is acquired, it is dispersed throughout the main collection and not kept as a separate entity.

The Collection at Work

Mineral specimens in the national collection are utilized in research conducted by members of the museum's staff as well as by scientists elsewhere. While much of this work relates to problems of identification, classification, and refinement of existing data on individual species, complete studies of interesting mineral environments are also undertaken. The Department of Mineral Sciences of the Natural History Museum is well equipped to perform such studies. It has one of the few classic wet-chemical analysis laboratories still in constant use today, a carry-over from the time when it was the only analytical technique available. Its importance, however, has never abated, even with the recent addition of a number of sophisticated modern analytical instruments, including the electron microprobe, the scanning electron microscope, and X-ray fluorescence and atomic absorption units. The museum's X-ray diffraction laboratory is similarly well equipped with a half-dozen different instruments. Few museums in the world can provide the technical research support found in the National Museum of Natural History.

Valuable specimens that are not on exhibit line shelves in the "blue room," where they are stored securely.

The World of Gemstones

The fiery brilliance of diamonds, the rainbow iridescence of opals, the rich gleam of rubies, the satiny translucence of jade—the magical mystery of gems has the power to delight and to beautify. Yet, these lovely stones are nothing more than modified minerals or rocks. In their rough state, most gemstones are not beautiful. Their full color and luster is revealed only by the skill of the lapidary who cuts, polishes, or engraves them.

Ideally a gemstone should be hard, durable, and impervious to changes in temperature, atmospheric pressure, sunlight, and common chemicals to which it may be exposed accidentally.

Gems display a wide range of hardness. On the Mohs scale, diamond, of course, is the hardest at 10. Next comes corundum (9), represented in the gem world by sapphire and ruby. Of the popular gem minerals, chrysoberyl is next (8.5), then topaz (8), followed, in order of decreasing hardness, by spinel, the garnets, beryl, tourmaline, quartz, spodumene, the feldspars, and opal, which is about 6 on the scale.

Hardness is not an absolute measure. Some gemstones, such as opal, can vary greatly. Other minerals display different hardness in different directions because of their internal structure. And members of such groups as garnets and feldspars also are not uniformly hard because they differ in composition. On the other hand, the gem varieties of quartz, which include amethyst, citrine, smoky quartz, and rock crystal, are identical in hardness.

In addition to being hard, a gemstone should be tough, particularly if it is set in a ring or bracelet, where it is exposed to more abuse than in a brooch or pendant. The toughest gems are the jades—nephrite and jadeite—and chalcedony. Although these stones have a hardness of only 7 or less, their structures, which consist of dense, tightly interlocking masses of microscopic crystals, give them great durability.

It might seem reasonable to restrict the use of the term "gemstone" to shaped and polished minerals and rocks used for personal adornment, but this would omit organic substances, including pearls, shells, amber, and coral, that have traditionally been considered gems, as well as unset gems, such as those often seen in museums and smaller private collections. And, of course, shaped and polished objects are at some point too large to be considered gemstones and are called sculpture instead. Clearly there are no precise limits to the term "gemstone," and a collection of gems may well include carvings, both small and large, and other objects such as spheres, boxes, and snuff bottles fashioned from minerals, rocks, and organic substances.

From observing the most antique gems, it seems apparent that the earliest preparation of gemstones involved little more than applying a polish to crystal faces. Later, pieces of minerals were rounded and polished to create beads and round or oval cabochons.

Perhaps just as early, artisans were carving stones and engraving images on the surfaces. Eventually, it was learned that a series of flat faces, or facets, could be ground on various sides of a small crystal piece. When polished, the stone reflected light, both within and without, making it sparkle like natural quartz crystals, which may well have inspired the first faceting. In time, artisans learned that certain angles for facets maximize the bril-

The Smithsonian's collection of gems includes many large stones of great beauty. Clockwise from upper right: a 610-carat Australian fluorite; a 354.4-carat Korean fluorite; a 99.7-carat South African fluorite; a 640-carat citrine and a 210-carat amethyst from Brazil; and an 815-carat topaz from the U.S.S.R..

This exquisite modern rendition of Leda and the swan was carved in Germany in banded, dyed agate.

THE WORLD OF GEMSTONES

The dazzling pendant of this diamond and gold necklace is a 68-carat, champagne-colored Victoria-Transvaal diamond, which was discovered in South Africa in 1951.

liance of a gem, and that these preferred angles are different for different types of gemstones.

All of these techniques are widely used today. Less transparent, more massive stones, often variously colored and intricately textured, usually show to best advantage when carved or rounded into cabochons. Even crystals of moderately transparent gems that are too extensively flawed for faceting look well when carved or engraved.

Faceting is usually reserved for very transparent, almost flawless, and preferably brilliant or sparkling crystals. Not surprisingly the most favored gems are generally those that are prettiest when faceted. The exceptions are the easily scratched or "soft" minerals, such as fluorite, which would not be durable set as jewelry, but can be valuable as exquisite carvings. Because faceted soft stones have little commercial value, they seldom appear in the market but can be seen in museums. There, protected by glass, these frequently elegant, rare, fragile gems can be displayed without risk.

Scarcity and public demand make a gem highly desirable. Historically, some long-recognized gems, including emeralds, sapphires, rubies, and diamonds, have been characterized as "precious." But most of them are no more precious than a number of other gems, some of which have been discovered recently. Only the very best emeralds, reasonably free of inclusions, with excellent color, can command top prices. Inferior emeralds, or sapphires, rubies, and even diamonds, are unlikely to be worth as much as tanzanite, tsavorite, and especially red beryl. As a designation then, "precious" is not valid; its continued use only further confuses the already frustrated consumer.

Scarcity, in the context of "precious" gems, refers to the amount of fine gem-quality stones available. The gems sapphire and ruby, for example, are both forms of corundum, a mineral so common that its nongem form has been widely used for making sandpaper because it is very hard. Only as a gem is corundum scarce; consequently, the very best rubies and sapphires may bring exceptionally high prices.

There is no great mystery about what makes a gem desirable. It must, first, be beautiful. It should also be uncommon, a term used in preference to "rare." Diamonds are not rare. Literally millions of carats of gem diamonds have been removed from the earth in the course of diamond mining, but most of the stones are small. Only a tiny percentage possesses the best color, which in the case of diamonds usually means the total absence of color, but may also mean a strong, attractive color like blue or pink. The 1/4-carat to 1/2-carat stones often set in engagement rings are relatively inexpensive because they are not rare. Less common, however, are colorless diamonds of large size, 1 carat or larger. These stones command high prices, which escalate rapidly with increasing gem size.

Surprisingly, a gem may be too rare to be commercially interesting. A few extremely valuable gems are in such short supply that they are known only to a relatively small group of private collectors. Since the consumer is not familiar with them, they are seldom used for jewelry and never appear in conventional retail outlets. Red beryl, or bixbite, from Utah is an excellent example. Others include taaffeite and kornerupine, whose total production in cut gems is readily absorbed by the collector market.

Gemstones have their own nomenclature, which has come into usage helter-skelter, with little concern for order or understanding. Many current names originated hundreds of years ago. Two of these, for example—ruby and sapphire—both refer to the mineral corundum. The mineral beryl may be known in the gem trade as aquamarine, emerald, morganite, heliodor, or even bixbite, depend-

MEASURING GEM WEIGHT

Around the end of the nineteenth century, jewelers settled upon the carat as a convenient measure of the weight of a gem and defined it as a weight of exactly 200 milligrams (0.2 gram). Most dealers in rough gem material still use the gram, but the conversion is easy since there are 5 carats to a gram. The weights of small diamonds are often given in points: there are 100 points in a carat. A 75-point diamond will weigh 0.75 carats. The term "carat" should not be confused with "karat," a measure of the weight proportion of gold in precious metals.

ing upon its color. The nomenclature of gems can be confusing to the uninitiated, but it is easily learned, for there are probably fewer than a hundred important names.

The world of gemstones is surrounded by an aura of mystery. But gemology is not a complex subject, the variety of gems is not extensive, and many exquisite and exciting gems are affordable.

Napoleon I gave this crown to his consort Empress Marie Louise. Set in silver, the 950 diamonds weigh 700 carats. The 79 original emeralds have been replaced with Persian turquoise cabochons.

Discovering Gemstones

Like ores, most gemstones are concealed beneath the earth's surface, and it is necessary to dig into the crust to recover them from rocks. But some river gravels hold concentrations of gem materials, while other deposits are exposed loose on deserts.

Recovering most gems, diamonds for example, involves mining. The famous Kimberley diamond mine in South Africa is the largest excavated crater in the world. Begun in 1871 and worked until 1914, it is 3,620 feet deep. Underground mines are not the only source of these gems. Diamonds are also found loose in gravel in the alluvial deposits of old riverbeds, and there are rich underwater deposits along the coast of South Africa where diamonds were left millions of years ago by ancient rivers. So diamonds, at least, are mined in a variety of ways.

By contrast, other gems are recovered, even today, by the most primitive form of mining—hand cobbing. Peridot, for example, is mined this way on the San Carlos Reservation in Arizona. Here Apache families work piles of volcanic rock, using picks and hammers to carefully break out "knots" of olive-green gem peridot tightly encased in the basalt. The deposit is very large and promises a continuous supply of peridot far into the future. Simple though the mining may be, it is estimated that the San Carlos Reservation is the source of 80 to 95 percent of the world production of peridot.

Elsewhere in the Southwest, lone prospectors track down thin veins of turquoise in outcroppings in the desert. These occurrences are usually small and may be completely mined out in a couple of days.

Between these extremes, gem mining takes almost every imaginable form. Sapphire deposits in sands and gravel from old riverbeds in Montana, for example, are first attacked with powerful hoses; streams of water break down the gravel so that it can be screened to separate out the gem corundum crystals. In Maine and California, underground tunnels are dug along circuitous courses in pegmatite rock in pursuit of the colorful gem tourmaline. To uncover the elusive, but greatly coveted, intense red beryl called bixbite, dense, white volcanic rock is blasted with dynamite in quarries on

Gem tourmaline is mined in Newry, Maine, using heavy equipment.

the flanks of the Wah Wah Mountains in Utah. In Oregon, at the Ponderosa mine, bulldozers push volcanic rocks around in the quest for sunstone, a beautiful orange-red feldspar with tiny inclusions of copper that reflect light, giving the gem a sheen called "schiller."

Most gemstone mining is small-scale because gem material is sparsely disseminated and gem crystals cannot be recovered by the rough methods used for ore mining. Gem crystals must be extracted without breaking them: the more perfect they are and the larger they are, the greater the profit. Untold fortunes have been lost through careless and inappropriate mining.

Throughout much of the history of gem production and sales, few people seemed to care where gems came from. Until quite recently, on the rare occasions when the place of origin was given for a gem, it was to establish that the gem represented the ultimate in quality. Gems from certain localities represent a standard of excellence with which all others are compared. A gem might, therefore, be labeled a Kashmir sapphire, a Siberian amethyst, or a Colombian emerald. Some, though not all, stones from these sources are universally recognized as the supreme examples of these gems.

Today, interest in knowing where a gem was mined seems to be rising. Some retail dealers take pains to make sure that the country of origin and often even the mine name are identified with gems for sale. In many cases, this information helps sell the stone. A Californian, for example, might be more inclined to purchase a tourmaline from California than one from Brazil.

Museums welcome this change in attitude because curators have always placed more emphasis on provenance than most consumers do. If there were not so many possible sources for fine gems, this information might be less important. Gem tourmaline, for example, is found all over the world, so it is virtually impossible to guess where a particular stone came from. This is true of many gems, including diamond, sapphire, amethyst, and most topaz. While it is easier to guess the source of rarer, more exotic gems, it is still just a guess. It is far better if this information is established at the outset.

Deposits of fine gems are scattered around the world, but some countries possess clusterings of exceptional richness. While the United States is endowed with a great variety of wonderful gems, especially tourmaline, amethyst, sapphire, garnet, spodumene, morganite, red beryl (bixbite), topaz, opal, feldspar (including sunstone), turquoise, and peridot, with the exception of peridot and red beryl it is far from being one of the world's major suppliers. The country that surely leads the rest is Brazil, which has a wealth of fine gems. Brazil produces a substantial share of the world's tourmaline, aquamarine, topaz, amethyst, citrine, spodumene, garnet, rose quartz, and agate. In addition, Brazil is also known for opal, emerald, diamond, and a large number of exotic, less well-known gems.

Southern Africa is still rich in diamonds, but some countries here are becoming important suppliers of sapphire, emerald, red garnet, and a pair of newer gems—tanzanite and tsavorite. Australia has long been known for the best opal, but is now also a major producer of diamonds, even rare pink ones. Madagascar and Mozambique have been important as well, as have India, Mexico, Sri Lanka, and the Soviet Union. Colombia is the dominant supplier of fine emeralds. More recently, Afghanistan and Pakistan have gained recognition for production of a variety of gems; especially notable is Afghanistan's near monopoly on top-quality lapis lazuli. China's deposits are only now being exploited and seem to have great potential.

Fortunately, gems are dispersed all over the world, assuring a good supply of wonderful stones far into the future. And many gem deposits, occurring as they often do in remote and inaccessible places, remain to be discovered.

On the San Carlos Reservation in Arizona, gem peridot is mined by hand.

How Gemstones Are Cut

A person who shapes and polishes gemstones is known as a lapidary, derived from the Latin word *lapis*, which means stone. A skillful lapidary can transform a rough, unattractive stone into a gem of great beauty. The techniques that the lapidary applies can be quite diverse and depend upon the nature of the stone and the desired result. Cutting cabochons, for example, is a very different process from faceting.

Cutting Cabochons

Cabochons, polished but unfaceted, are almost always round or oval—shapes chosen to best enhance interesting textures or colorful markings of opaque or translucent stones. The stone is usually ground into the desired shape using a vertical abrasive wheel over which flows a stream of water to lubricate and cool both the wheel and the gemstone. After the gem is shaped, it is finished on a succession of finer and finer wheels, or flat disks, until it is free of scratches. A final polish is then applied with a soft buffer and a polishing substance, such as zinc oxide powder, mixed with water. The basic process is not complicated, but like any other craft, cutting cabochons requires special skills and extensive practice.

With most gemstones, choosing the orientation of a cabochon is easy because it merely involves selecting the surface pattern or desired appearance of the finished piece. Star stones and most cat's-eyes, however, must be carefully oriented because the inclusions that make them so special are directional. Cut in the wrong direction, the stars will not show to best advantage or even at all.

Faceting

Faceting is preferred for almost all transparent stones. Even before beginning to shape a stone, the lapidary must visualize the completed gem. If the surface of the raw piece is rough, it may be necessary to polish a "window" on it in order to look for internal flaws and to determine the presence and

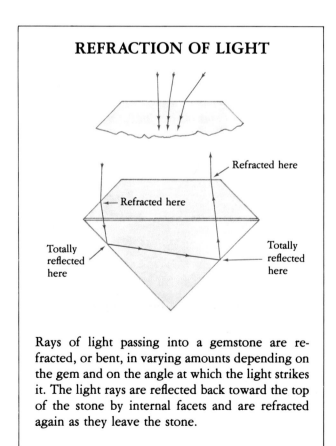

REFRACTION OF LIGHT

Rays of light passing into a gemstone are refracted, or bent, in varying amounts depending on the gem and on the angle at which the light strikes it. The light rays are reflected back toward the top of the stone by internal facets and are refracted again as they leave the stone.

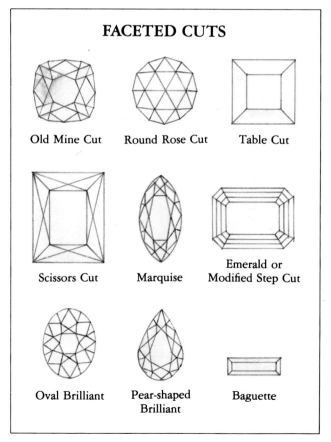

FACETED CUTS

Old Mine Cut   Round Rose Cut   Table Cut

Scissors Cut   Marquise   Emerald or Modified Step Cut

Oval Brilliant   Pear-shaped Brilliant   Baguette

orientation of irregularities in color. A knowledge of crystal optics is essential, for the optical properties of a gem greatly influence what may be done with it to bring out the best it has to offer.

Some stones are more "brilliant" than others. A brilliant gem diffracts more light into spectral colors and, at the same time, appears more highly reflective, not unlike a bright mirror. Brilliance can be enhanced or diminished by the stone's cut. In some gems the property of brilliance is directional, so orientation is important when the stone is cut.

To maximize the potential of most gems, facets must be positioned in specific, angular relationships. A well-cut stone causes light entering from the top surface, the table, to reflect off the bottom surfaces, the pavilion, and back out through the table. With this ideal cut, a stone becomes as brilliant as its optical characteristics allow it to be.

Coloring within most gems is usually splotchy or banded because of zoning in the crystal. Poor orientation of a gem will not disguise the lack of uniform coloring. Proper orientation, on the other hand, causes the color to appear evenly dispersed throughout the gem, when it is viewed from the table.

Some gems are pleochroic: that is, they absorb light differently in different directions. For this reason, a pleochroic gem is one color in a certain direction, but a different color in another. Pleochroism may affect only the intensity of color, or it may change the hue as the stone is turned. This, too, must be kept in mind when deciding which orientation is best.

To begin faceting, the lapidary determines the rough form of the gem, shaping it with a diamond saw, a thin metal disk edged with tiny diamond chips. Then the preformed stone is mounted on a rod fitted into a faceting device that allows the lapidary to position it at any angle. After the angle is set, the stone is lowered onto a horizontal grinding wheel and a facet is ground or "cut." The stone is repositioned and a new facet is cut. The lapidary continues until all facets of half of the gem are cut, and then repeats the process with a fine wheel to polish them.

Finally, the lapidary turns the stone around and remounts it on the rod so the facets of the other half can be done. The "halves" are usually not identical because the facets of the crown are usually cut in a pattern unlike that of the pavilion, especially for stones to be set in rings, pendants, or pins. The lapidary carefully plans all differences to make the stone as striking as it can possibly be.

The aquamarine set in this superb yellow gold, platinum, and onyx pendant is an example of Bernd Munsteiner's innovative fantasy cut.

Styles of Cutting

Lapidaries are always looking for new and notably different designs. Bored with conventional cuts, they strive for artistic variation. In the early 1980s, Bernd Munsteiner, a third-generation German lapidary, created what has been dubbed the fantasy cut. This completely new style of free-form cutting elegantly marries sculpture with faceting in unconventional geometric shapes interrupted by bold diagonal slashes. Today, Munsteiner is widely acclaimed and, predictably, has many imitators. His most important contribution, however, may have been to create an atmosphere conducive to innovative, imaginative work, truly stimulating the gem cutter's art.

Diamond

With its exceptional luster and fiery brilliance, diamond is unquestionably the king of gems. In most parts of the world, no gemstone commands more attention and respect. More mysteries and fantastic legends surround the diamond than all other gems combined. The notion that nations have fought over diamonds, that empires have been toppled, and the knowledge that men have killed for them, add immeasurably to the diamond's appeal.

Diamonds are the gems of choice for those who want to display their wealth, yet some of the world's great diamonds have disappeared, either hidden in someone's vault or recut beyond recognition. The histories of many diamonds have been wildly exaggerated or hopelessly confused. Frequently, facts about one diamond have been attributed to another.

Although the diamond is best known in the Western world as the stone set in the ring traditionally given to celebrate an engagement of marriage, this practice probably did not begin until Victorian times. It was not until production began in South African mines, after 1870, and major technical developments were introduced in the art of cutting at about the same time, that diamonds were put within reach of almost everyone.

Diamond, a form of pure carbon, is not only the hardest gem, it is the hardest mineral and, in fact, the hardest substance known. But diamond is not indestructible. It is brittle and can shatter on impact, and it will readily burn. Some diamonds are even lost in cutting because the process exacerbates internal stresses and can cause the stone to fly apart.

Cutting a Diamond

Because of its unparalleled hardness, it is difficult to imagine how a diamond is "cut." A popular image pictures a diamond cutter cleaving a diamond by tapping upon an edged tool positioned on the stone's surface to make it break in a certain direction. Cleaving is, indeed, sometimes used, but because it is effective only in four directions, it is mostly used to split large stones or to remove nongem portions of a crystal.

Cutting a diamond actually involves sawing and

The brilliant cut, left, was developed to maximize the brilliance and fire of diamond. The beauty of both the 16.7-carat Pearson diamond, left above, and the 5.03-carat DeYoung diamond, above, is enhanced by the brilliant cut. The Pearson diamond is from South Africa. The origin of the rare red DeYoung diamond is unknown.

In 1811, Napoleon I gave this 275-carat diamond necklace to Empress Marie Louise, to celebrate the birth of their son, the future king of Rome.

grinding, just as does faceting other gems. In the case of diamond, the cutting agent must be diamond. For sawing, tiny diamond pieces are set in the edge of a circular blade. For grinding and polishing, a paste of diamond dust is applied to a revolving disk. The hardness of diamond varies with direction; it is measurably harder in the octahedral plane than in the cubic plane. Therefore, facets ground onto a diamond cannot lie in a plane of the octahedron: it would be impossible to produce or polish them.

A diamond is brilliant because it has a very high refractive index (a measure of the degree to which light is bent and dispersed when it enters a transparent medium), because its polished surfaces are highly reflective, and because it is commonly cut in a style that maximizes these features—the "brilliant" cut. In a fine diamond, light entering the stone is dispersed into spectral colors, described as the gem's "fire." The absolute absence of internal color and imperfections further enhances a diamond's beauty and brilliance, but large perfect stones are exceedingly rare and exceptionally valuable.

Colored Diamonds

Collectors and consumers consider it quite acceptable for a diamond to have color, but only if the color is sufficiently deep. Pale color, especially light yellow or brown, is considered a detriment and greatly diminishes the value of the stone. Pronounced blues, pinks, and yellows are positive attributes for marketing. These diamonds are referred to as "fancy-colored" and they are rare, especially blue and pink stones.

The Smithsonian's 2.9-carat DeYoung pink diamond was found in Tanzania and is generally regarded as one of the loveliest known. In recent years, a large diamond deposit in Australia has proved to be the world's first important source of pink diamonds, but the stones are rather small and are lighter in color than some collectors prefer.

Corundum

Gem corundum is found in almost every conceivable color, all called sapphire unless the color is red, in which case the gem is ruby. Since corundum is second only to diamond in hardness, both sapphires and rubies take a brilliant polish that greatly enhances their beauty.

Rubies

It is sometimes difficult to determine what is and what is not ruby. While ruby's color has traditionally been described as "pigeon-blood" red, there is no universal agreement about exactly what that color is. Rubies are among the most expensive gems, so some dealers would like to stretch the description of ruby to include many stones that might well be called pink sapphires.

The ultimate but seldom seen ruby is a rich red stone of perfect clarity. Stones that are not perfectly transparent lose some brilliance and are described as "sleepy," a derogatory term that translates to being worth less money. If the color of a ruby is too light or has too much of a violet component, the proper designation is pink sapphire. Fine rubies are coveted and bring extraordinary prices. To add further to their allure, cut and polished gems of top quality come only in relatively small sizes, so a stone of a few carats is considered an extremely good one. Rubies of 10 or more carats are exceptionally large.

Many collectors consider that the most superb rubies are mined only in the Union of Myanmar (formerly Burma). Almost all the remaining sources of rubies are located in Asia. Thailand boasts the greatest production; Sri Lanka, India, and Pakistan supply modest amounts of mostly less desirable grades. Small deposits of ruby are also found in South Africa.

Sapphires

When a stone is called a sapphire it usually means that it is blue. Sapphires of other colors are customarily preceded by the name of that color, and they occur in an incredible range, including pink, violet, yellow, green, and "white," or colorless. Blue sapphires are by far the most plentiful, especially today when inferior stones can be treated and turned a good blue. (This treatment is usually limited to small stones because it can be readily detected in larger ones and would greatly diminish their value.) Since sapphire is the same mineral as ruby, it displays similar characteristics.

Blue sapphires from Kashmir have always been the standard with which all others are compared, but the area has not produced any new stones for a very long time. The best stones today also come from the Union of Myanmar. Although Sri Lanka is more prolific, its gems tend to be light in color. Australia is a major supplier of small stones, and Montana has long been a source of blue sapphires as well as sapphires of other colors.

One sapphire in particular deserves special mention. The padparadscha must rank at or near the top of the list of most desired exotic gems. Named for the color of the lotus flower, this reddish-yellow orange gem corundum is so rare that few examples of it are available for comparison. There is, therefore, widespread disagreement about how best to describe it. Sri Lanka is to date the only source of true padparadscha sapphire.

Cutting Rubies and Sapphires

Most gems dictate the manner in which they are cut to best advantage. Because of the intensity of color of deep sapphires and rubies, they gain little from light being directed into them and reflected back out. Their great asset is their color, so they are usually cut in shallow oval shapes that allow more light to be transmitted from beneath. Other, paler gems benefit from deeper cuts and better manipulation of incident light through cutting to precise angles.

Star Rubies and Sapphires

Imperfect stones of ruby and sapphire frequently contain silky "veils" that detract from their beauty and value. A clever cutter can disguise these imperfections, just as patchy color resulting from zoning can be homogenized, making the gem appear much better than it really is.

These fine sapphires, from Sri Lanka, exemplify some of the more subtle colors of this gem.

The sapphires in this diamond and sapphire necklace come from Sri Lanka and weigh 195 carats.

Although inclusions are often considered flaws, with several gems—and corundum in particular—more than a little of an inclusion may be a good thing. It is common for corundum crystals to have extensive inclusions of hairlike needles of the mineral rutile (titanium oxide). Typically, these inclusions are aligned along any of three geometrically equivalent directions within the crystal. Consequently, a correctly oriented cabochon made from such a crystal will produce a six-rayed star when a beam of light is pointed toward it. Stones of this type are called stars, whether rubies or sapphires. Although other gems also display stars, star rubies and sapphires are the most popular, and the better ones are certainly the most valuable.

Synthetic Rubies and Sapphires

Corundum was one of the first gem crystals successfully grown in the laboratory. The process has been perfected so that it is possible to grow ruby and sapphire crystals much more perfect and far larger than those found in nature. Since these crystals are not natural, and can potentially be made in infinite amounts, they do not really rival the natural gems, at least in terms of the mystique of rarity. While laboratory-made crystals are frequently cut into gems, they are usually identified as synthetic and are, therefore, modestly priced.

Scientists have also learned to make rubies and sapphires with rutile inclusions, making synthetic star stones available. They are easy to spot because the stars are far more perfect than nature's.

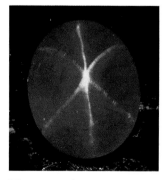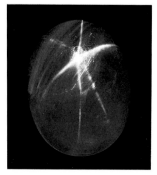

The extraordinary 138.7-carat Rosser Reeves star ruby is from Sri Lanka. The Star of Bombay, bequeathed by Mary Pickford, is a remarkable 182-carat star sapphire.

The Beryl Family

Aquamarine, emerald, heliodor, and morganite are varieties of the mineral beryl. These gems differ only in color caused by chemical impurities so minor that they do not change the general composition of the mineral.

The beryl family has been extensively studied. It is known that trace amounts of particular elements impart distinctive colors to the various gems. Iron produces the blue or light-green color of aquamarine; chromium (or, rarely, vanadium) gives emerald its velvety green. Iron makes heliodor golden-yellow, and manganese imparts the pink to morganite and the intense hot pink or red to bixbite. Still another, little-known beryl gem, goshenite, is colorless.

Aquamarines

By far the most abundant of the beryls, aquamarines may be found in immense, glassy, gem-quality crystals that weigh as much as several pounds. Brazil provides the world with most of its aquamarine. Prodigious quantities of fine crystals have been found there since the initial discovery in the seventeenth century.

Blue is most the desired color of aquamarine. Stones that are green, a common natural color of Brazilian beryls, are often heated to remove the yellow component, leaving them blue. Heat treatment of blue stones also tends to intensify or darken their color. It is assumed that virtually all blue aquamarines sold on the market have been heated. Because aquamarine crystals are often large and almost flawless, it is possible to cut rather remarkable gems of unusually impressive sizes.

Emeralds

Although also green, like some aquamarines, emeralds are colored by chromium rather than iron. They form under very different circumstances than aquamarines, and emerald crystals are rarely large. Even the typically small ones are seldom flawless. Consequently, the flaws are part of their lore—an emerald is frequently described as containing a "garden." Small emeralds are routinely sold with imperfections that would never be tolerated in aquamarines of any size. Emeralds, too, may be "enhanced" by heating, in this case in certain oils that permeate any microscopic cracks, making the defects less visible.

Colombia has had a near monopoly on the production of gem emerald for so long that the name of the country has become synonymous with the gem. Although Colombia's emerald deposits were exploited even before the Spanish arrived in the sixteenth century, they are so vast that there is little chance that they will be exhausted in the foreseeable future. Very fine emeralds are found in limited amounts elsewhere, in particular in Pakistan, Tanzania, Zimbabwe, and Brazil, but the total combined production of these countries and all others is insignificant when compared with Colombia's output.

There are many varieties of beryl. Above is a fine 911-carat Brazilian aquamarine; right, a crystal of the red beryl, bixbite, from the Wah Wah Mountains of Utah.

The brownish yellow, 330-carat morganite gemstone, below, and the 235.5-carat, pink morganite gemstone, right, are from Brazil.

The Smithsonian's collection includes a number of exceptional emeralds. Particularly notable is the 168-carat Colombian pendant of the Anna Case Mackay necklace. It may be the largest fine-gem emerald that is set in a piece of jewelry.

Morganite

Morganite was named in 1910 by G. F. Kunz, the noted Tiffany gemologist, to honor the financier J. P. Morgan, who was, it seems, one of Kunz's best customers. It is not known if Kunz was simply trying to curry favor with Morgan or if he thought that the name would help to gain public acceptance for the pink gem, which was not then widely recognized. Whatever the objective, the effort paid dividends; J. P. Morgan eventually donated his excellent collection of gems to the American Museum of Natural History in New York City.

Brazil and Madagascar have long supplied much of the gem morganite, but California and Maine also contribute to the supply. A mine in Maine has only recently begun producing excellent cutting material, although the color is pale.

Newly mined specimens of morganite are peach to brownish-orange. After prolonged exposure to the sun, the color changes to pink.

Heliodor and Goshenite

Heliodor is golden-yellow, a color that is not popular. Relatively large stones of 30 to 50 carats are available because yellow beryl is found in fine, large, unflawed crystals in deposits similar to those of aquamarine. Goshenite, which is colorless, is also unappreciated, so there is not much of a market for the moderate-sized stones that are cut. Although they are beautiful, both heliodor and goshenite are curiosities snapped up by museums rather than by important jewelers. Since these gems are not actively marketed, they tend to be modestly priced.

Bixbite

The red beryl, bixbite, is rare. The only locality in the world known for this gem material is a mine in the Wah Wah Mountains of Utah. The crystals are always rather small, and seldom is more than a tiny portion of a crystal suitable for cutting. Very small stones of less than a carat are, therefore, the norm.

Because of its rarity, bixbite is virtually unknown to the public. This gem would gain immediate popular acceptance if it were available in sufficient quantity. Mining is difficult (huge volumes of rock must be moved for every crystal recovered), but production has recently begun to increase.

Opal

In composition, opal is similar to quartz. Both are silicon and oxygen, but opal typically contains some water. At one time, opal was thought, like glass, to lack a regular atomic structure of the kind present in quartz. More recent studies have revealed that opal actually consists of microscopic spheres with a disordered crystal structure resembling that of cristobalite, another form of silicon and oxygen. Water may occupy interstices between the spheres; although not essential, the amount of water present usually varies from 3 to 10 percent.

The play of color that makes opal so beautiful is due to the manner in which stacked layers of spheres diffract light. Different colors are produced by stackings of groups of spheres of different diameters, and the colors produced in a single opal vary when the angle at which light strikes the piece is changed. This is why several different colors can be observed in a single place on an opal as it is moved around in the light.

Common opal is relatively abundant and is of no value as a gem because it displays no play of color and is not attractive. Precious opal used for gems is of many types. The patches of vivid color seen in precious opals may be limited to two or three colors, perhaps red and blue, or they may include a broader spectrum, especially yellow and green with red and blue.

Stones whose color is imposed on a milky translucent body color are known as milk or white opals; those with color patches on a dark-gray to black background are called black opals. When angular blocks of color patches appear quiltlike and the overall base color is dark, the stone is known as a harlequin opal. The size of color patches can vary, too, from large irregular squares to the tiny but brilliant dots of so-called pinfire opal. Jelly opal ranges from translucent to transparent, and the colors are usually quite pale.

Mexico is the exclusive source of fire opal, so named because its body color is orange to orange-red or amber, and not because of its color play. The term "fire" must be used carefully when applied to a gem. With a diamond it refers to the dispersion of light and the spectral colors generated in an often colorless stone. It is sometimes applied to the small pin spots of brilliant color in opal. But "fire opal" is quite specific to the reddish or amber Mexican opal, with or without color play. Fire opal is usually found in little nut-sized "knots" in volcanic rock, so it seldom appears as large gems. A type of jelly opal, fire opal is often called cherry opal by Mexicans. It seems to be widely distributed throughout Mexico.

Most experts rate black opal as best, surely because the dark background provides a pleasing contrast to the vivid colors of fine opal. The best of the black opals are from an area called Lightning Ridge in New South Wales, Australia. Virtually all gem black opal comes from Australia, as does most white opal. Many years ago, some exceptional black opal was found in Nevada, but almost all of it proved to be unstable: in time it dehydrated and cracked, destroying the beauty of the pieces.

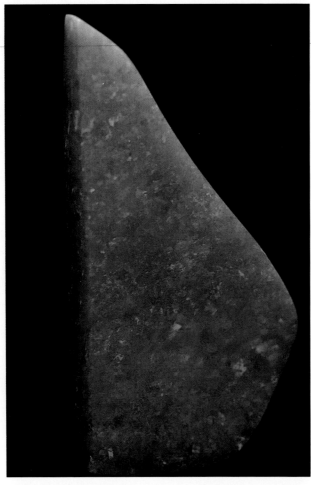

In this Australian white opal, tiny points of color are dispersed within the white matrix.

The Dark Jubilee Opal, above, is a 318.4-carat, free-form stone. Technically a black opal, it comes from Australia.

While many opal deposits occur in volcanic rocks, those of the vast Coober Pedy deposit in South Australia are in sandstone, a sedimentary rock. Veins of opal-bearing material tend to be small, and the best gem opal is found in thin seams. Much of the mining is done by hand. Dynamite cannot be used because opal is extremely fragile; a blast might shatter all gem opal in the immediate area.

Opals have traditionally been cut as cabochons. Since they are never perfectly transparent and the desirable color effects are generated at or near the surface, this is the only sensible form. Often the cutter leaves some of the matrix on the back of a finished gem to give it more body and strength.

Opal must be treated respectfully. Not only is it fragile, it is one of the softer major gems and much of it is prone to cracking if it is overheated or simply dries out. Sudden temperature changes can also destroy it. All of these shortcomings, however, do not diminish the opal's inherent beauty that makes it a popular gemstone.

The 29.85-carat fire opal below was mined in Jalisco, Mexico. Because of its unusual transparency, it was faceted rather than cut into a cabochon.

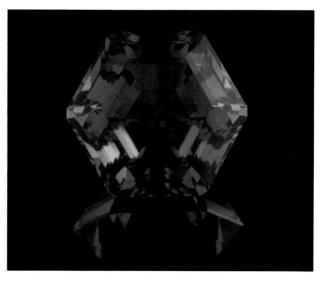

Quartz

In its many forms, colors, varieties, and possible uses, quartz is the most diverse natural substance known. The great bulk of quartz is plain and uninteresting with little or no specific value beyond its utility as sand or gravel. Nevertheless, when compared to other gems, gem-quality quartz abounds and large or even very large crystals are plentiful.

Many of the gem varieties of quartz have been given definitive, universally recognized names. Fortunately, however, not all varieties have their own names because a glossary of them would be overwhelming. As it is, people are more or less free to coin their own names when they find a type of quartz that is sufficiently different from all others. Because of the casualness that characterizes naming quartz varieties, there are hundreds, each with a peculiar name chosen to honor a person associated with it or to commemorate the place where it was discovered.

Among the generally recognized quartz gems are amethyst, citrine, smoky quartz, and rose quartz. While the names amethyst and citrine are usually used alone, the term "quartz" is always used with a modifier like "smoky." Colorless, or transparent, quartz is called rock crystal, an intriguing name because it derives from the belief, originating at least as early as the Middle Ages, that quartz crystals were permanently frozen ice. To add to the confusion, the term "crystal" by itself refers almost exclusively to man-made glass.

Rock Crystal

Large, colorless quartz crystals are plentiful throughout much of the area around Hot Springs, Arkansas. Rock crystal is also particularly common in clefts in the rocks of the Swiss, French, and Italian Alps. In Herkimer County, New York, countless little, clear quartz crystals fill cavities, or pockets, in sandstone; because of their sparkle and beauty, they have long been known as Herkimer diamonds. In contrast, Cape May diamonds are small, rounded pebbles of quartz, found in the sand at Cape May, New Jersey. Although frosted

From Brazil, this highly faceted rock crystal egg is 7,000 carats. Most of the small faceted stones in the supporting stand are Montana sapphires.

There are many varieties of quartz gems. They include, from the left above, smoky quartz, rock crystal, amethyst, citrine, and rose quartz. The smoky quartz is from Switzerland, the other stones from Brazil.

This 24-carat ametrine, from Bolivia, was oriented by the lapidary to produce a bicolored stone.

The lovely amethyst crystals right, come from Due West, South Carolina.

by weathering, they are perfectly transparent and can be faceted into small stones.

Rock crystal is seldom faceted because quartz is not particularly brilliant, and a finished stone looks no better than cut glass. It can, however, be carved into elegant forms. Although it still resembles cut glass, carved rock crystal is prized because it is much less common than glass and is more difficult to carve because it is harder than glass.

Amethyst

Amethyst is purplish- to violet-colored quartz, the best of which may show flashes of red. When it does, the darker stones are called Siberian. Examination of an amethyst crystal reveals colored zones in pie shapes or thin bands alternating with colorless ones, which are typically parallel to crystal faces. It is difficult to imagine from its rough state that a crystal could produce a large gem amethyst in which the color appears evenly distributed, yet this is what happens when a gem is properly cut. Quite large amethysts are not at all rare, and the supply of small ones is seemingly endless. Differences in quality, which may seem subtle to the untrained eye, can make profound differences in value, especially in larger stones.

Citrine

Citrine is yellow to yellow-orange or yellow-brown quartz. While yellow quartz actually occurs naturally, most of the citrine set in jewelry is heat-treated amethyst probably from the geode beds of southern Brazil and northern Uruguay. Citrine is not especially popular and does not command high prices, in part because of its reputation for having been treated and also because even the naturally colored gems are not particularly attractive.

Ametrine

Since the early 1980s, a remarkable new gem quartz has been mined in Bolivia. It is composed of zones of natural citrine dispersed throughout amethyst. Unusual and striking color effects can be achieved when an inventive lapidary cuts this gemstone. It is possible to produce a bicolored gem, one part distinctly yellow-orange and the other amethyst, or the stone can be oriented to blend the two colors. While several names have been proposed, ametrine seems to be catching on as the preferred name in the trade for this quartz.

Smoky Quartz

Smoky quartz can grow into extremely large, uniformly colored crystals. It is sometimes cut into stones big enough for doorknobs and paperweights, but it is not particularly attractive. (Some dealers intentionally mislabel it "smoky topaz," a

In this dramatic specimen from Crystal Peak, Colorado, long, tapering, smoky quartz crystals emerge from the blue-green microcline crystals with which they have grown.

Rose quartz crystals encircle smoky quartz in this magnificent specimen from the Sapucaia mine in Minas Gerais, Brazil.

practice that is disappointingly widespread.) Smoky quartz does lend itself admirably to carving.

Rose Quartz

Rose quartz is not very popular as a faceted gem because its pink color is often too pale and the stones are always cloudy. When rose quartz is cut in cabochons, or rounded into beads for necklaces, or carved, it is far more effective.

Inclusions

Large glass-clear crystals of quartz will grow around and over existing crystals of other minerals. These inclusions can be beautiful, especially when they are slender golden needles of the mineral rutile (titanium oxide). Inclusions can also be used to advantage in creating exquisite finished gems and carvings.

Like corundum, quartz is hexagonal; that is, it has sixfold symmetry in one direction. Therefore, when it contains tiny silky needles of rutile, it may display six-rayed stars, as corundum does. This phenomenon is especially common with rose quartz, whose color is also due to finely dispersed crystals of rutile.

Aventurine, a granular quartz rock that contains tiny green crystals of a mica densely dispersed throughout, is used extensively for beads.

Chalcedony

There are many types of chalcedony, a cryptocrystalline variety of quartz composed of microscopic crystals. Under extreme magnification the crystals can be seen to consist of aggregates, or bundles, of tightly interlocking fibers. These fibers give chalcedony a distinctive toughness. When chalcedony is broken, the fresh surface appears waxy, not glassy, as other transparent quartz crystals do.

The most familiar gem forms of chalcedony are carnelian, which is red; sard, which is darker to brownish red; and chrysoprase, which is bright green. Chalcedony that has replaced asbestos is known as tiger's-eye. Plasma is dark-green chalcedony colored by inclusions of actinolite. When plasma contains red spots of iron-oxide "rust," it is called bloodstone.

All agate is chalcedony. Found in numberless forms and diverse colors, agate can be highly translucent, almost transparent, but it is always at least a little cloudy. Whites and grays are common colors, but any impurity can tint agate so the range of colors is virtually unlimited. Most authorities feel that chalcedony must be banded to be considered ag-

Polished agate cabochons display interesting patterns. The one on the left is from Idaho, the other from Washington.

The pattern in this dendritic agate, from India, was formed by crystals of a manganese mineral.

ate, but since there are all gradations between agate and ordinary chalcedony, the distinction becomes fuzzy.

Onyx is banded agate, alternating between black and white, or red and white in the case of sardonyx. Much of the banded agate that is marketed has been artificially dyed. Since alternate layers of agate often have very different porosities, dyes are absorbed by some layers and not others. Regrettably, calcite and aragonite are also referred to as onyx. These forms of calcium carbonate can be quickly distinguished from quartz because they are easily scratched and will dissolve in a weak acid like vinegar.

More translucent types of agate that contain tree-like growths, usually of a black manganese oxide mineral, are called dendritic or moss agate. Those with light feathery inclusions are called plume agate. Despite the names, most authorities prefer to consider these stones chalcedony rather than agate because they are not banded.

Agate is frequently associated with volcanic rocks. Residual heat from lava is a factor both in increasing the amount of silicon dissolved in hot waters and in precipitating that silicon with oxygen in the form of agate. Round gas bubbles in volcanic rock provide excellent spots for agate formation. In old lava flows there are often concentrations of egg-sized to fist-sized agate-filled nodules called thunder eggs. The name is said to be of Native American origin. When thunder eggs are sawn open, they may reveal attractive bands of concentric layers of alternating colors, or a star pattern. Nodules that never filled completely with agate are known as geodes; these hollow nodules tend to be lined with quartz crystals.

Other forms of chalcedony, including jasper, chert, and even flint, have limited gem applications. Some jasper boulders reveal spectacular patterns when cut open. The western United States has long been an excellent source of jasper, and it is not surprising that varieties have been given names like Bruneau Canyon, Biggs, flower, fish-egg, and poppy.

All three of these forms of quartz—jasper, chert, and flint—have played an important nongem role in the past. Because of their tendency to form smooth concave surfaces when they break, they were ideal materials for flaking into arrowheads and spearheads for primitive weapons. Flint, of course, was used to produce the spark to ignite gunpowder in muskets and other early firearms.

Topaz

Because it is hard, topaz is a good stone for rings. It is also one of the more affordable faceted gems and has long been popular. The extraordinary abundance of gem-quality stones can be accounted for because topaz not only occurs in deposits in many countries but also tends to be found in huge crystals.

The most successfully marketed colors are the deep-orange to red-orange stones from Ouro Preto, Brazil. Known as precious or imperial topaz, they are recovered in moderate quantities from a vast deposit, so it may be assumed that the stones will be more or less available for a long time to come.

Much rarer from Brazil, and also found in very limited amounts in Pakistan, is pink topaz. This highly prized gem is certainly the most expensive topaz on the market. It was frequently used in fine Victorian jewelry.

Sherry-colored crystals are found in Mexico; they also occur in Utah, but shortly after exposure to daylight become colorless. Natural blues come from the Soviet Union, Brazil, Nigeria, and Colorado.

Most natural topaz is colorless, but since a colorless topaz has about as much appeal as glass, it has remained more a curiosity than anything else until recently. In the early 1970s, blue topaz, produced

There is great color variation in topaz crystals. The 3,273-carat Brazilian topaz above has a wonderful natural soft blue color. The 22,892.5-carat Brazilian topaz left is known as the American Golden. It was presented to the museum by the American Federation of Mineralogical Societies and is one of the world's largest gemstones.

This interesting spray of sherry-colored topaz crystals is from the Thomas Range in Utah.

by irradiating colorless topaz, began to appear on the market. Today the volume of blue topaz produced through irradiation is measured in millions of carats sold annually. These stones have become a popular substitute for very rare and expensive aquamarine of a comparable blue. The induced blue is permanent and is more intense than natural blue topaz, which is always quite pale. The artificial stones are actually more attractive, except for the deep inky blue shades, which look too unnatural.

In 1988, the Smithsonian was given what was then the largest faceted gemstone known, a golden yellow topaz from Brazil, weighing 22,892.5 carats. Its moment of glory lasted for no more than a year. It was soon surpassed in size by an even larger if less attractive topaz, also from Brazil. That there are three faceted topazes in excess of 20,000 carats, including one in the American Museum of Natural History in New York, attests to the available wealth of facet-grade topaz of large size.

The Garnet Group

Garnet is a group name for certain silicate minerals that share a similar structure. The garnet minerals —almandine, pyrope, spessartine, grossular, andradite, and uvarovite—are all closely related in chemical composition. Each contains impurities of one or more of the metals of the other garnet minerals. For example, an almandine, which is mainly iron aluminum silicate, may also contain varying amounts of magnesium and manganese, making it part pyrope and spessartine, too. The species name given to the mineral is that of the dominant component. In this example, the mineral is called almandine because it contains more iron than magnesium or manganese.

The first gem garnets used in jewelry, either large oval-shaped cabochon "carbuncles" or tiny faceted stones, were probably the deep red ones found abundantly in central Europe, India, and elsewhere. Their color is so intense that, except in strong light, cut stones of any size appear almost black. Consequently, they were often cut in a rather bizarre way. The backs of the stones were made concave, scooped out so that the stones would not be too thick to allow light to pass through.

Most of these early stones were certainly the mineral almandine. Other red garnets used for gems, especially pyrope, can vary in color from deep red to violet-red. Gem pyrope is now called rhodolite. Undoubtedly, some spessartine also found its way into Victorian jewelry, but large deposits of this garnet were not discovered until well into the twentieth century.

Today, almost all of the red garnets that are mined in large quantities and extensively used in jewelry are much more attractively colored than the dark red antique stones. Darker garnets are still used, but primarily as very small beads for neck-

Garnets occur in many colors. Above are a 9-carat orange grossular garnet from Sri Lanka, a 10-carat green demantoid from Russia, and, of the more traditional red color, a 74-carat Tanzanian rhodolite and a 109-carat Brazilian spessartine.

laces; in this size they appear red rather than black.

In the past, only one other garnet was attractive enough to be used much as a gem—the grass-green to emerald-green variety of andradite, which is called demantoid. The best crystals were always very small and so rare that demantoid was never widely known. Limited amounts good enough to cut have been found in Russia and Italy.

From the middle of the twentieth century onward, the garnet picture has changed dramatically. Extensive mining of gem pegmatites, especially in Brazil, has produced lots of spessartine. The Jeffrey asbestos mine in Quebec has turned up a considerable amount of the gem grossular, which comes in light-orange to pink shades and has been cut successfully. Newer and quite large deposits, mainly in Tanzania, have made rhodolite a far more prominent gem garnet than ever before.

A green grossular, colored by traces of vanadium, was discovered in Kenya and Tanzania in 1968 and named tsavorite, after the Tsavo game preserve. The limited quantities of tsavorite produced so far have been well received because of the intense, emeraldlike color of the best of the stones. Gem collectors who are well aware of more unusual, but highly desirable, exotic gems like tsavorite have taken to it with enthusiasm because it is as lovely as emerald but not as expensive. Tsavorite is also found in lighter and yellower shades, which are less highly valued.

A wonderful new gem garnet has also been found in Tanzania in fine crystals large enough to generate cut stones of exceptional size that are not too dark. This garnet is difficult to name because its composition varies over a substantial range between pyrope and spessartine. It is, perhaps, the loveliest of the red garnets because its color has strong orange or peach and pink undertones. In spite of its beauty, the rough gem was scorned when first discovered and was given the name malaya, which in Swahili means "outcast" or "prostitute." Ever since the gem's beauty has been acknowledged, the name has not been too widely used; many dealers prefer to call it simply "garnet."

Some garnets also have silky inclusions, and the best of them will display stars when properly cut. Since garnets are isometric they possess high symmetry, essentially that of a cube. Consequently, star garnets can be oriented to produce either a four-rayed or a six-rayed star, depending upon the axis of rotation. The best star garnets are found in Idaho.

Almandine garnets are often found in schist. This specimen is from Alaska.

This orange grossular garnet crystal was found in an asbestos mine in Quebec, Canada.

Victorian jewelry, like this hair ornament, was often set with dark red pyrope garnets from Czechoslovakia.

Jade

Jade is the traditional name for two distinctly different materials—jadeite and nephrite. Jadeite is one of a group of silicate minerals called pyroxenes. Nephrite may refer to several minerals of the amphibole group.

Technically, both jadeite and nephrite are rocks. Therefore, they differ from gems that are single crystals or, at least, a single substance, like opal and the fine-grained forms of quartz known collectively as chalcedony. Like other rocks, the jades are masses of tightly interlocking, mostly microscopic crystals. Jadeite may have crystals of other minerals, usually albite, scattered through it. Nephrite also can contain many different minerals.

Nephrite is similar in texture to chalcedony. Its dense, tightly interlocked, microscopic fibrous crystals give the rock a toughness that is characteristic of jade and make it an excellent material for intricate, detailed carving. The poorer grades of nephrite may have a grain, causing the stones to break irregularly.

Jadeite is more coarsely crystalline than nephrite; individual crystals may often be seen without magnification. Jadeite is almost as dense and tough as nephrite, however, and is eminently suitable for carving.

Nephrite can often be distinguished from jadeite by the appearance of the polished surface. Nephrite tends to be greasy or waxy because it does not take as high a polish as jadeite. This is not an absolute test, however, because a piece of poorly polished jadeite could be confused with nephrite.

Neither jadeite nor nephrite is particularly hard, but both are about as tough as steel. Since they are often found in huge masses, they are frequently carved into large ornate objects. Green nephrite was used for large as well as small intricate carvings in ancient China, where jade carving was developed to a fine art. Pre-Columbian civilizations of Central America carved jadeite, and for several centuries the Maoris of New Zealand have used nephrite for ornaments and ceremonial objects.

This graceful jadeite bowl was carved in China in the eighteenth century.

Today, jadeite and nephrite are extensively used for jewelry, especially as cabochons for rings, as carved bracelets, and as beads.

Nephrite and jadeite range from translucent to almost opaque, with a wide color variation. Although green is usually associated with jade, it is only one of its many colors. Nephrite is usually a dark shade of forest green, but it can be lighter or yellowish-green, or even creamy white to light gray. Jadeite is more variable. It is frequently white; when the white is streaked with veins of apple green and even lavender, it is particularly beautiful and highly coveted. Other colors of jade include red, orange, yellow, brown, and blue-gray.

The most prized jadeite is a translucent, deep emerald-green stone that takes a brilliant polish, which gives it a bright finish. This is imperial jade, and the best qualities have uniform color. Imperial jade is rare and correspondingly valuable. Unlike other gemstones, the value of the more common, less beautifully colored jade carvings is predicated as much, or more, on the beauty and antiquity of the carving as on anything else, including the color. A simple cabochon of the finest quality imperial jade can, however, exceed in value most of the oldest and most skillfully carved jade pieces.

Boulders of nephrite and jadeite are often found in streambeds, and the outside surface may be altered through weathering. This brown outer skin is often cleverly worked into the design of a carving with pleasing effects.

Jades are found in many areas throughout the world, and most of the jade that was not found in China was sent there for cutting. It is doubtful that China ever produced any jadeite. The Union of Myanmar (formerly Burma) has always been the source of the finest jadeite. It was first brought to China from Burma in the middle of the eighteenth century. The poorer-quality Central American jadeite has been found in substantial amounts in Guatemala. Nephrite exists in noteworthy amounts in New Zealand, western Canada, Taiwan, Wyoming, and Alaska.

Many substances are carved in Asian styles and passed off as jade. The commonest is serpentine, a much softer mineral. Since it is softer and can be scratched with a knife blade it can be easily identified, and since it is more easily worked than jade, details of the serpentine carving are usually finer. Other jade substitutes include aventurine, chrysoprase, dyed chalcedony, dyed Mexican onyx, and soapstone. With a little experience, all of these substances can be easily distinguished from true jade.

The Dragon Vase above is carved of rare lavender jadeite from Myanmar. A modern piece, it is 18 inches tall.

The pair of altar lanterns below were carved of nephrite for the Chinese Emperor Ch'ien Lung in about 1750.

Lesser Known Stones

Either because they are rare or only recently discovered, many gems remain relatively unknown. Some of the rarer gems have been around since ancient times, yet are still not widely known, and probably never will be, for there is so little of each of them that they seldom appear in conventional retail outlets. Gems that have been used for centuries and whose names appear often in older written references are assured of widespread recognition, but those of twentieth-century finds, regardless of how wonderful they may be, do not fare as well. For example, few people may have ever heard of sinhalite, taaffeite, benitoite, brazilianite, or even tanzanite. And were it not for museums, few people could even hope to see most of these exciting "new" stones.

Between the rare and the familiar are many gems whose names are at least vaguely familiar. Peridot, tourmaline, spinel, zircon, alexandrite, chrysoberyl, cordierite, feldspar (either orthoclase or labradorite), kyanite, scapolite, and kunzite have been known to the trade for a long time, but these gems have never enjoyed general recognition.

Peridot

Peridot is the gem name for a yellow-green mineral that occurs in volcanic rock; the mineral's species name is forsterite. Tiny crystals of peridot have been found in recent lava flows on the island of Hawaii, but are too small for gemstones. Most of the world's supply is mined in Arizona, hand-hammered out of basalt.

A transparent gem, peridot has a distinctive oily luster. The color ranges from pale golden-green to brownish green. The most valuable stones are a rich deep green. Today, peridot is most commonly seen in irregularly shaped, tumble-polished pieces that are drilled and strung for necklaces.

Tourmaline

Tourmaline is actually the name for a large group of minerals, most of which are moderately abundant. As with garnet, in the jewelry trade the name tourmaline typically refers to any gems of this group, regardless of the species.

Although they form in many geological environments, most tourmalines are coal black and virtually useless as gemstones. Relatively large crystals of gem-quality tourmaline in a great range of color, mostly shades of red and green, occur in many pegmatites, the coarsely crystallized rocks produced by residual fluids exuded by large igneous intrusions.

The proper mineral name for most gem tourmaline is elbaite, named for the island of Elba, where this species of tourmaline was found long ago. Although there are many sources of gem tourmaline worldwide, two of the most famous are in the United States. One is near Rumford, Maine, and the other around Pala, California. Both have had long histories of production, beginning with the discovery of gem tourmaline near Paris, Maine, in

The 310-carat, fine-quality peridot, from Zebirget, Egypt, is the largest known. The 8.9-carat peridot is from Arizona.

Exceptionally fine color distinguishes this 48.7-carat green tourmaline from Newry, Maine.

The various colors in this natural candelabrum of elbaite, a mineral of the tourmaline group, were produced by changes in the solutions from which the crystals grew. It was found at the Tourmaline Queen mine in Pala, California.

1820. Seventy years later, the gemologist George F. Kunz described the early Maine tourmaline as rivaling any in the world. California was not known as a source of gem tourmaline until 1870, but easily surpassed Maine in production. Early in the twentieth century, single mines turned out literally tons of crystals. Maine's production has always been much smaller, with the exception of a remarkable find at the Dunton mine in Newry. Mining in 1972 and 1973 yielded at least several tons of beautiful tourmaline, much of it eventually cut into red or green gems.

While mining in both areas continues, it is unlikely that the total combined production approaches that of Brazil, unquestionably the world's greatest source of gem tourmaline. The first Brazilian gem tourmalines made their way to Europe in the eighteenth century, but were not well received, either because the stones were of poor quality or were not cut well. Following the excitement of the discovery of beautiful tourmaline in California, the gem gained wider acceptance, and mining in Brazil really began. New discoveries are continually being made there, and some exceptional ones have had a notable impact on the market in recent years.

Spinel

Spinel is probably better known for its mistaken identity as something else, usually ruby, than for its considerable merits as a gemstone. It is not surpris-

Spinel is found in a broad range of attractive colors. Above, a rosy pink 22-carat gem from Sri Lanka, a ruby red stone from Myanmar, and a steely blue spinel from Sri Lanka.

ing that red spinel has been confused with ruby and also with garnet; its properties and appearance fall more or less between the two.

Spinel is best known in any of a variety of shades of red, but it is also found in many other colors. A rare and highly prized cobalt-bearing variety of fine sapphire-blue color is found in Sri Lanka. Spinel, an isometric mineral, typically occurs in simple octahedral crystals that are sometimes left uncut, polished in their natural forms, and effectively used in jewelry.

Zircon

The colorless varieties of zircon are so brilliant that this gemstone was once widely used as a substitute for diamond. Consequently, its name has come to be associated with fake diamond, and the gem is often regarded with suspicion. Zircon can be beautiful and is found in many colors, of which the brilliant light blue is possibly the best. A lapidary must take great care when faceting a zircon because the stone can be lifeless and dull if it is not cut properly.

Caution should be exercised when selecting zircon for a ring because it has a reputation for chipping easily. Zircon is not the same as cubic zirconia, a synthetic of different composition, used extensively today as a diamond simulant.

Alexandrite

This gem has the enchanting property of changing color from leaf green in daylight to raspberry red under incandescent light. Alexandrite is the only one of a number of rare "color-change" gems marketed in any quantity. The first alexandrite gems were alleged to have been found in the Ural Mountains of Russia in 1830, on the birthday of Czar Alexander II, for whom the stone was named. It has since been found in Brazil, Zimbabwe, and Sri Lanka, but it is not plentiful, and stones as large as 5 carats with a strong color change are difficult to find. Alexandrite is a gem variety of the mineral chrysoberyl.

Chrysoberyl

In addition to alexandrite, chrysoberyl is known as a gem in two other forms. It is more abundant in larger yellow to yellow-green crystals than as alexandrite. These stones can be cut into large gems of 20 to 30 carats or more, but do not show a color change. Brilliant pale yellow-green chrysoberyl stones from Brazil were popular in eighteenth-century Spain and Portugal, where they were used in jewelry.

Large chrysoberyl crystals may contain inclusions of exceedingly fine parallel needles of another mineral or hollow tubes, both called "silk." When cut in cabochons, these crystals produce a cat's-eye from light reflected by the silk. Although many other minerals may display this effect, the best and most traded gem cat's-eyes are chrysoberyl.

Cordierite and Kyanite

A sapphirelike gem found principally in Sri Lanka and India, cordierite (also known as iolite) is really a collector's stone because there is so little of it. It has a pleasing blue color and is strongly dichroic (blue in one direction, yellow to violet-gray in the others). It is noticeably softer than sapphire.

Kyanite, another rich blue gem, also bears certain similarities to sapphire. It, too, is primarily a collector's stone. Clean gems are rare, although large blue crystals are found in Brazil, Kenya, and North Carolina. Kyanite has the unusual characteristic of varying substantially in hardness depending upon the direction in which the property is mea-

Chrysoberyl is abundant in large yellow to yellow-green crystals. The specimen far left is from Espirito Santo, Brazil. The handsome 114.3-carat chrysoberyl gem left is also from Brazil.

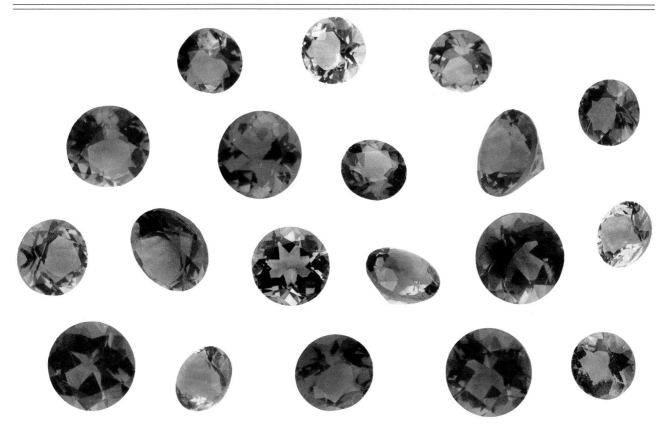

Sunstone, a gem variety of labradorite, is found in warm salmon to rich red crystals with a golden sheen. These gems came from the Ponderosa mine in Oregon.

sured. It is 7.5 in one direction and only 4 or 5 in another, undoubtedly contributing to its reputation for being difficult to cut.

Cordierite and kyanite could easily be confused with sapphires but they seldom are because these rare gems are almost never sold commercially.

The Feldspar Group

There are a number of gems in the feldspar group, but few have ever gained prominence. That may soon change because a deposit of labradorite in Oregon has the potential to produce enough of the gem sunstone to make it a household name.

Labradorite is a feldspar species. Sunstone is the name of a gem variety of labradorite with two distinctive characteristics: it is found in warm salmon to rich red crystals that also have a golden sheen, because of microscopic platelets of bright copper aligned in parallel planes dictated by the crystal's structure. Sunstone has been known for many years, but only recently was it discovered in substantial quantities, in Oregon.

Moonstone, another gem of significant antiquity, is also in the feldspar group. The origin of moonstone is quite different from that of sunstone. When the feldspar crystals of moonstone first formed, they were rich in both potassium and calcium. Upon cooling, the crystal structure adjusted to lower temperatures and these components became mutually incompatible, so they "unmixed," or separated into two different feldspar minerals, one of potassium and the other of calcium. They became alternating platelets of microscopic thickness within the crystal. Light diffracted from their surfaces produces the beautiful soft sheen or bright iridescence of moonstone.

Orthoclase, a potassium feldspar, is found in

This intricately carved moonstone from Sri Lanka is about one inch in diameter.

LESSER KNOWN STONES

This rare lavender scapolite from Tanzania is 8.1 carats.

One of the largest cut kunzites, the 880-carat gem above, from Brazil, is almost flawless. Below is a spodumene crystal like that from which gem kunzite is cut. It was found in the Vandenberg mine near Pala, California.

large glassy crystals in Madagascar. These crystals have produced modest numbers of very attractive gems of good size, but the yellow stones are little known beyond the world of gemstone collectors.

Scapolite

There are strong similarities between gems of the feldspar group and those of scapolite, which is also a group, although a limited one. Scapolite gem crystals are of good size and vary from colorless to yellow and amber, or more rarely violet and pink. Like the feldspars, they are relatively soft (for gemstones), and they have very similar optical properties. Scapolite is primarily a collector's gem and virtually unknown in conventional gem markets.

Tanzanite

Although the mineral species zoisite has been established for nearly two centuries, it was not until the early 1960s that this gem variety of the mineral was discovered in Tanzania. Tanzanite, a blue sapphirelike gem, is moderately abundant. It was actively promoted by Tiffany's, which, in 1969, named it in honor of the country in which it was discovered. Nevertheless, tanzanite is not as well known as might be expected.

Spodumene

The mineral spodumene is best known as a gem as kunzite, named after the famed gemologist George F. Kunz. This variety is an exquisite violet-pink. Often found in remarkably large and perfect crystals, especially in Brazil, Afghanistan, and California, kunzite frequently yields oversized gemstones. Unlike many deeper-colored gems that must be cut into smaller stones to avoid appearing too dark, kunzite works well as large stones, where its color is intensified.

Spodumene occurs in pegmatites and may be associated with tourmaline in many of them. Spodumene is also found in light shades of colors other than violet-pink—particularly blue-green and yellow—but these crystals are not considered kunzite.

Small, rich green spodumene crystals containing chromium are known as hiddenite. Both the gem and the town in North Carolina where it was found are named for W. E. Hidden, a gemologist who made the initial discovery while employed by Tiffany's. Hiddenite is truly a collector's-only gem because few faceted gems exist, all quite small. The largest faceted hiddenite in the Smithsonian collection is only 0.7 carat.

Massive and Decorative Stones

Gem materials unsuitable for faceting are often cut into cabochons or carved. This is especially true of larger pieces of nontransparent minerals and rocks. Relatively common and less valuable gem crystals and flawed crystals of some of the more valuable gems may also be carved.

Jade qualifies as a massive gem material, as do many substances that are widely utilized in carving figures and ornamental objects like bowls, spheres, vases, and boxes with hinged lids. Some of the better known examples of these materials are serpentine (a jade look-alike), malachite, gypsum (alabaster), lapis lazuli, rhodonite, sodalite, and talc. On a smaller scale, turquoise and variscite, both fine-grained and often interestingly veined by other materials, are popular for cabochons or for carving into small objects like snuff bottles.

Malachite, gypsum, rhodonite, and sodalite are actually species names, but they are also almost universally used by lapidarists to refer to rocks that are composed entirely or mostly of one of these mineral components. Because they may also contain impurities in the form of crystals of other minerals, many of these rocks display variations of color in interesting patterns, making them highly desirable for carving.

Malachite

Malachite's wonderful concentric swirls of light and dark green are related to the size of the crystals in the alternating bands. Zones with the tiniest crystals are lighter than those in which the crystals are coarser. Patterns in malachite are revealed by making cuts across copper carbonate stalactites or stalagmites.

An immense deposit of massive banded malachite in Zaire, in the province of Shaba, is the source of almost all malachite used for contemporary carvings and jewelry.

Turquoise

Turquoise forms when solutions carrying phosphorus and copper move along fractures in aluminum-rich rocks. If conditions favor a reaction between these solutions and the rocks, a fine-grained, powder-blue vein of turquoise precipitates, eventually filling the available space in the fracture. Tur-

Malachite, a copper mineral with swirling patterns of bright and dark green, is carved as well as used for jewelry. This specimen is from Zaire.

quoise seams, or stringers, are usually small, so most discoveries produce only limited quantities of gem material.

Pueblo Indians mined turquoise in the American Southwest long before the Spanish arrived. At the Los Cerrillos mine in New Mexico, trees almost seven hundred years old grow on its dumps. And at Pueblo Bonito, New Mexico, archaeologists uncovered about thirty thousand turquoise beads and pendants in just one room of a burial site. The dating of relics from these early mines indicates that turquoise mining goes back at least to the fifth cen-

This specimen, from Los Cerrillos, New Mexico, shows veins of turquoise as they formed in the original matrix.

The charming turquoise perfume bottle above is intricately carved in the Chinese style.

tury. Turquoise mining in Persia (now Iran) probably began even earlier.

While many people regard Persian turquoise as the best, because of its uniform, rich sky-blue color, others prefer the less perfect American turquoise whose patterns of black or brown veining are called "spiderwebbing." Some American turquoise is decidedly green and considered less valuable than the blue. Turquoise is relatively soft (5 to 6 on the Mohs scale) and porous. If it is not protected with an impregnation of wax, it may absorb body oils and cosmetics and eventually become green.

Turquoise may be carved when pieces are large enough, but most of it is cut into cabochons. A popular and effective use of small angular pieces is to cement them together to create a mosaic which, in turn, is shaped into cabochons.

Lapis Lazuli

Lapis lazuli is usually referred to simply as lapis, but its mineral name is lazurite. Most of the stones seen in the market are really rocks with irregular splotches veined with calcite and pyrite. Nevertheless, the whole mixture is legitimately called lapis. Lapis has been mined for more than six thousand years. For many centuries the only known deposits were in Mongolia and at Sor-e-Sang, a remote valley in Afghanistan. From here lapis found its way to Egypt and the ancient kingdom of Babylonia and, later, was traded into Europe. More recently, Chile has produced substantial amounts of lapis.

The lazurite component from both Mongolia and Chile is a less intense azure than the lazurite from Afghanistan, which is found in very large, nearly pure masses or nut-sized single crystals, easily removed from the sugary white marble in which it formed. Although Afghan lapis has been mined for thousands of years, its quality has been extraordinary in more recent times.

Lapis is used in carving, in making boxes, as cabochons for jewelry, and as beads for necklaces.

Chalcedony

Many forms of chalcedony, especially agate, are extensively used in all kinds of decorative objects. Banded agate, tiger's-eye, chrysoprase, chert, jasper, bloodstone, onyx, petrified wood, and chrysocolla (chalcedony in which the blue copper mineral chrysocolla is dispersed) are generally abundant, relatively inexpensive, and well suited to such treatment.

Fluorite

One of the most attractive and usually transparent carving materials is fluorite. Not only is it often found in vivid colors, especially deep purple or green, but it tends to appear in exceptionally large crystals or layered veins, both of which may typically show sharp bandings, thin layers that alternate abruptly from intense to light colors or even colorless layers. Fluorite is a very soft mineral (4 on the Mohs scale) and is so brittle that objects carved from it have little utility beyond decoration. Much of the earliest carving of fluorite was done with the deep-blue banded vein material from Derbyshire, England, which became famous as Blue John. Vases and other decorative objects were made from this variety of fluorite more than fifteen hundred years ago.

Charoite

As is true with the faceted gems, some of the best lapidary materials have been discovered in modern times. Charoite, a rich lilac-colored mineral, was discovered about 1978 in the Soviet Union. The rock that is composed mostly of charoite goes by the same name for lapidary purposes, although other minerals—orange, black, and light gray—are intermixed. The source of charoite must be extensive; great blocks of this rock have appeared for sale. It is usually fashioned into cups, vases, and bowls, but decorative square tiles have also been available, and it has occasionally been used in jewelry.

Sugilite

Another new gem material, sugilite is remarkably similar to charoite. It, too, is recovered in rock form in which the mineral sugilite is the dominant component. Originally found in Japan and considered a nondescript curiosity, deep lavender to reddish-violet gem sugilite was discovered in 1975 in a manganese mine in South Africa. The purest sugilite is quite translucent and is used primarily for jewelry. Larger pieces of lesser quality are usually carved; the natural patterns of the rock make the figures and ornamental objects particularly attractive. The South African deposit appears to be exhausted, but experts feel that more gem sugilite may await discovery underground.

These pieces are carved of mottled purple charoite from the Chary River in Yakutsk in the Soviet Union.

Organic Gems

Amber, coral, shell, pearl, jet, and ivory—known collectively as organic gems—are derived from plants and animals. They have been used for personal adornment since the Stone Age. Beads of amber, ivory, and shell have been found at Paleolithic grave sites and ruins.

Amber

Amber is the general name for a large group of fossilized tree resins that vary greatly in their chemical and physical nature. Most ambers are very light and will float in seawater. Amber was one of the earliest materials used as gems because it was found lying in profusion along the shores of the Baltic Sea and was easy to collect. It was also quite easily worked into beads, since it has a hardness of only about 2 to 2.5 on the Mohs scale. Amber is found in many colors, but it is predominantly yellow with shades of orange, brown, or even red.

Today, most commercial amber comes from the Baltic coasts of Russia and Poland, with lesser amounts from the Dominican Republic, Sicily, Myanmar (formerly Burma), North America, and Mexico. Caution must be exercised when buying amber because fakes, especially plastics, abound.

Of great interest to science are ambers that contain perfectly preserved insects, pollen, and even tiny animals—victims caught in the sticky sap of living trees millions of years ago.

Coral

Gem coral is an outer skeletal framework built by tiny, warm-sea animals capable of extracting calcium carbonate from seawater and secreting it as calcite, the mineral of which most coral is composed. Entire reefs of coral encircle many islands in the South Pacific.

Corals typically grow in branching treelike forms, and they can be variously colored. Most coral is fashioned into beads and cabochons, but finer pieces, especially those of a rich salmon color, are often beautifully carved.

While coral is usually white, shades of pink and even red come from the Mediterranean and the South China Sea. Black coral, which comes from Hawaii, is a completely different substance—a fibrous protein material called conchiolin.

Like all organic gems, coral is soft and therefore fragile. It must be protected from abrasion, and all except black coral will dissolve in any liquid that is even weakly acidic.

Shell

Appealing in shape and color, seashells are also a source of gem material. Mother-of-pearl, which has traditionally been used for buttons and inlay work, is the iridescent lining of mollusk shells. Larger shells provide mother-of-pearl thick enough for carvings and beads.

Giant conch and helmet shells consist of layers of

Within this piece of amber are tiny insects and a feather, perfectly preserved.

Carved of fine salmon-colored coral, the fish is only 2½ inches long. In the foreground are branches of natural coral.

contrasting colors; these very large shells are often carved into cameos. Conch-shell cameos have white images on a pink background, while helmet shells yield white images on a brownish background. Because shell is soft and easy to work, the carving may be elaborately detailed.

Pearls

Both freshwater and saltwater mollusks, particularly oysters and mussels, create most of the world's pearls. Pearls are produced wherever these animals can live, although areas where commercial quantities are generated are limited and, sadly, are becoming even more so because pollution is taking a toll on mollusk populations.

The subtle, iridescent luster of pearls is the result of diffraction of light by microscopically thin, concentric layers of translucent aragonite, called nacre, secreted by an animal around a bit of foreign matter, perhaps a grain of sand or a parasite, within the shell. This soft, silky luster is superimposed over a body color, which can vary from white to cream, rose, gray or silver, bronze, or even blue-green. The body color is influenced by the presence of varying amounts of conchiolin, the fibrous protein material of which black coral is also composed.

Small, round pearls matched in color and graduated or matched in size are prized for necklaces. Large, irregularly shaped baroque pearls are effective when combined with other gems in innovative jewelry, especially pins and rings.

Since demand far exceeds the supply of natural pearls, especially those that can be matched in size and color, pearls are now grown, or cultured. Several hundred years ago, the Chinese discovered that "seeds" (usually tiny spheres made of clam shells) could be implanted in the glands of living mollusks. The animals would coat these foreign objects with nacre, creating completely natural-looking pearls that could be harvested in several years. In the early twentieth century, the Japanese began mass-producing cultured pearls, which are readily accepted in the marketplace and, predictably, are less expensive than natural pearls. The culturing of pearls has remained almost exclusively a Japanese industry.

Jet

A form of coal, jet is easily carved and takes a good polish, despite its softness. It has been used to make decorative ornaments since the Bronze Age and was particularly popular during the reign of Queen Victoria, when it was used for mourning jewelry. It is rarely used in modern jewelry.

Ivory

Most contemporary ivory is elephant tusks, but it may also be walrus tusks or whale teeth. Obviously, collecting ivory usually requires that a great beast be killed. Used in jewelry, ivory was once highly fashionable and noncontroversial because much of the earlier material was fossil ivory that came from the ancient mammoth, an extinct member of the elephant family. Today, as elephants, too, are threatened with extinction, trading in ivory is widely regarded as distasteful and is illegal in some countries.

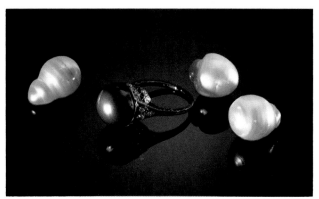

The three lustrous cultured baroque pearls above come from Japan. The superb cultured black pearl is from California.

The elegant, finely detailed cameo, left, is carved in a helmet shell from the West Indies.

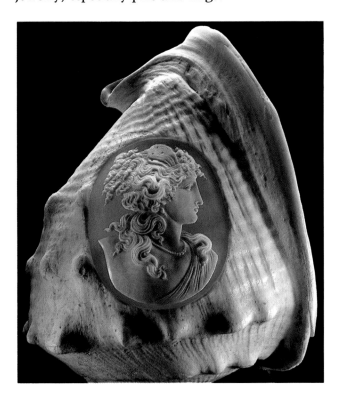

Synthetic, Imitation, and Assembled Gems

For thousands of years, gemstones have been imitated in many different ways. Methods of producing gems have been perfected in recent years, and, in many cases, it is nearly impossible for all but a highly trained gemologist to distinguish between natural and man-made gemstones.

Synthetic Gems

The first gem-quality synthetics were rubies, produced in the first years of the twentieth century by Auguste Verneuil, a French chemist who invented the flame-fusion process that is still used, with refinements, to manufacture rubies, sapphires, and spinels. Today, crystals of sapphire and ruby, emerald, spinel, and alexandrite can be grown in the laboratory. These gems are vastly superior in size and perfection to those found in nature. In all important properties, laboratory-grown gems are virtually indistinguishable from natural stones.

A synthetic gem, however, never rivals a natural gem in value, even though it may surpass it in quality. Fortunately, gemologists can still recognize gems that have come from the laboratory, but in some cases the task is extremely difficult because it is now possible to reproduce features that make crystals appear more natural and to eliminate features that would identify them as synthetic. It is, of course, considered highly unethical to try to pass off a synthetic gem as natural.

Although scientists have the ability to grow crystals of just about any known mineral substance, it is unlikely that synthetics of all the gemstones will be produced on a large scale; many of the processes are quite expensive and the value of the synthetic gems does not offset the cost of their production. Many of the crystals that have been synthesized to date were grown not for use as gems but because it was hoped they would have important industrial applications.

Imitation Gems

An imitation gem may look like a valuable natural gem, but it is not the same substance. Colored glass is probably the oldest and most widely used imitation because it is so easy to make. Paste, a brilliant glass with a high lead content, has long been used to fake diamonds. Today, many other materials come closer than paste to mimicking one or more of diamond's important properties, but anyone with even modest training in gemology can still easily recognize these substitutes.

Nontransparent gems may also be imitated. Very fine confetti-like platelets of various colors in glass create a remarkably realistic-looking opal imitation. Clay can be dyed blue, mixed with a binder, and then baked to create a false turquoise.

Dying natural materials is also widespread. Imi-

This remarkable, large, synthetic gem-quality ruby was grown by the Linde Corporation, a subsidiary of Union Carbide.

These are stones in some of the colors available in laboratory-grown yttrium-aluminum-oxide crystals. They have a garnet structure and are known in the trade as YAGs.

tation lapis is produced by treating chalcedony with blue dyes; other fake lapis is simply opaque blue glass. Glass with very fine silky inclusions closely resembles chrysoberyl cat's-eyes. Mexican onyx, which is a carbonate rock, is dyed a variety of colors. The green may be sold as Mexican jade.

Assembled Stones

Doublets are the most common of another kind of gem fakery. These assembled stones have been created since Roman times. Two parts are glued together to produce a stone that may be faceted or cut as a cabochon. Typically, the top portion is actually a gem, while the bottom is a different material—usually either glass or a colorless synthetic sapphire if the upper stone is too dark, or a colored synthetic sapphire if the top is light and in need of enhancement. The plane where the two are joined is usually positioned where the stone is to be mounted so that the setting obscures it.

A triplet, obviously, has three parts. The top and bottom may be pale-colored natural crystals of little gem value, such as near-colorless beryl. Glued in between, however, a slender wafer of intensely colored glass or synthetic sapphire imparts a rich color to the top. If either the top or the bottom of such a construction is tested, of course, it will prove to be a natural material.

* * *

Since fakes are common and are sometimes difficult to detect, it is wise to exercise caution when buying gemstones and carved objects. Perhaps the best rule, as when purchasing anything of value, is to buy only from reputable dealers. On the other hand, synthetics make it possible to own a fine emerald, ruby, or sapphire at a relatively low price. The only difference is that these synthetic gems don't have the magic and mystery of stones that evolved through complex geological processes over hundreds of thousands or even millions of years.

The National Gem Collection

The National Gem Collection had its beginning in December 1883 when Frank W. Clarke, a chemist with the U.S. Geological Survey, was appointed honorary curator for mineralogy. Although Clarke evidently had extensive experience with minerals, it seems that he had little knowledge of gems. Nevertheless, he prepared an exhibit of gems for the New Orleans Exhibition during the first year of his appointment.

The Commission for the Exposition appropriated $2,500, and with this Clarke was able to assemble a collection of one thousand specimens, which, according to the 1884 Annual Report of the Smithsonian Institution, consisted of "all the gems proper, rock crystal, agates and jaspers, malachite, lapis lazuli, etc., and every important gem ornamental species . . . secured both in the rough and cut conditions." When the exhibit came home to Washington, D.C., in 1886, it was displayed in what was then the newest Smithsonian facility, the Arts and Industries Building.

Although Clarke and George F. Kunz, the noted American gemologist, were good friends, the two major gem collections put together by Kunz did not come to the Smithsonian but were sold instead to the banker and financier J. P. Morgan, who donated them to the American Museum of Natural History in his home city of New York. That gift dropped the fledgling collection of the Smithsonian to second place in the nation.

Clarke was not deterred, however. In 1891, he again obtained money for the acquisition of gems for an exposition, this time the Columbian Exposition to be held in Chicago in 1893. With $500 he was able to purchase the collection of Dr. Joseph Leidy of Philadelphia. Catalog records reveal that the Leidy collection consisted of one hundred and fifty gems or groups of gems, an acquisition that helped greatly to recover some of the lost prestige.

This purchase was followed by the bequest the following year of a portion of the Isaac Lea collection, which consisted of 1,316 specimens of gems. Lea, like Leidy, was a Philadelphian. This bequest was followed in 1896, 1897, 1898, and even later by gifts of more gems from Isaac Lea's son-in-law, Dr. Leander T. Chamberlain. Most of Dr. Chamberlain's gifts were American gems. In 1897, Chamberlain was named honorary custodian for gems and precious stones. He served in this capacity until his death twenty years later, when he left an endowment of $25,000 to the gem collection. At that time the entire collection became known as the Isaac Lea Collection.

In 1910, when construction of the Natural History Building was almost completed, the gem collection was moved into it. There the gems exhibit changed little, but the quality and variety of specimens in the collection improved gradually, until both the gem and the mineral exhibits were renovated in 1958.

In 1921, the gem collection consisted of some four thousand individual stones, including "not only those used for personal adornment but as well such as are used in the smaller works of art and [for] utilitarian purposes." The collection was by no means a great one, but it did boast some exceptional items. Among them were a 58.5-carat green tourmaline from Mt. Mica, Maine, a 61-carat yellow orthoclase from Madagascar, and a 155.5-carat blue topaz from the Russian Urals. The topaz was a gem from the Leidy collection; the other two stones were purchased with funds from the Chamberlain endowment.

Early in the century, one of the collection's most extraordinary acquisitions was the donation by Nina Lea, granddaughter of Isaac and also a Philadelphian, of a set of sixteen faceted, yellowish-green sphenes (also called titanite), matched in color and graduated in size from 9.33 down to 0.76 carats. According to letters relating to the gift, Isaac Lea purchased the set in 1913 from the Duke of Devonshire collection, probably in London, and a value of $2,000 at that time was mentioned. While the source of the gems was never recorded, there is reason to believe that the stones may have come from Switzerland. Sphene is a wonderful gem with a brilliancy that rivals that of diamond. Although it is, unfortunately, too soft for most jewelry use, this does not diminish the stunning impact of this unique group of stones.

The Roebling collection, received in 1927, included an exceptional gem—a 7.6-carat benitoite

The Spanish Inquisition necklace, parts of which were made more than three centuries ago, is one of the many treasures in the Smithsonian collection. The emeralds are from Colombia, the diamonds from India.

from San Benito County, California. The Bulletin of the University of California of 1909 described this gem as being about three times larger than the largest flawless benitoite "so far obtained." Even today it may be the largest fine gem of its kind.

In 1928, the museum used $1,500 of the Chamberlain endowment to purchase a 65.7-carat alexandrite from Sri Lanka, an enormous stone for one of such excellent quality. It is still the largest alexandrite in the museum's collection and is undoubtedly one of the largest known, especially when alexandrites of more than 5 carats are considered extremely rare.

As plans for a new gem and mineral hall were underway in 1957, the stage was set for the generation of new interest in the growth of the collection. Museum officials discussed with Harry Winston of New York the possible acquisition of the famous Hope diamond. There was still sufficient time to incorporate a special vault for it into the design for the hall. When the hall officially opened on August 1, 1958, a huge safe was in place awaiting the arrival of the diamond.

The opening of the new hall, combined with the excitement surrounding the donation of the Hope diamond just one hundred days later, created a flood of publicity. For the first time, the public became aware that the National Gem Collection was well worth visiting. As a consequence, the museum began to receive impressive donations of gems and jewelry at a rate of more than one a year. Important additions to the National Gem Collection included the Inquisition necklace, the Chalk emerald, the Star of Bombay, the Dark Jubilee opal, the Anna Case Mackay emerald necklace, and the DeYoung red diamond, given as recently as 1988.

The growth of the National Gem Collection was phenomenal and, possibly, unparalleled. In a very short period, the Smithsonian's gem collection grew from relative mediocrity to one of great stature, quite likely the best publicly displayed gem collection in the world.

Hope Diamond

This most famous 45 1/2-carat blue diamond was formally presented to the Smithsonian on November 10, 1958, by Harry Winston. Not only is the Hope the largest faceted blue diamond in the world, it has the longest and most vivid history of any major diamond—the perfect complement to its mysterious icy blue color.

The diamond's recorded history began when Jean Baptiste Tavernier, a French gem merchant, bought a 112 3/16-carat blue diamond, already cut, in India, which he sold, in 1669, to Louis XIV of France. The king had it recut in 1673, reducing the diamond's weight to 67 1/8 carats. The recut stone has been described as heart-shaped, but a more accurate description would be a modified triangle with one very flat side. By all accounts, the gem was an intense steely blue and the royal inventories referred to it as the French Blue.

Louis XV had the stone set in a piece of ceremonial jewelry known as the Royal Order of the Golden Fleece. It was stolen during the French Revolution in 1792, along with several other important gems, including the crown jewels. The French Blue disappeared forever, but exactly twenty years and two days later, there appeared reports of a 45 1/2-carat blue diamond in London. Many pieces of evidence indicate that the gem was subsequently owned for some years by George IV of England. Between his death in 1830, however, and the appearance of an entry for it in the gem collection catalog of Henry Philip Hope, for whom the diamond is now named, there are no indications of its whereabouts.

Following Hope's death in 1840, the diamond changed hands a number of times before it was purchased in 1911 by Evalyn Walsh McLean of Washington, D.C. In 1949, Harry Winston bought the diamond from the McLean estate.

Marie Antoinette Earrings

Few objects in the Smithsonian collections conjure up more dramatic images than do these earrings. They were given to Marie Antoinette by Louis XVI and are said to have been taken from her when she was arrested as she attempted to flee from Paris during the French Revolution. This story, however, is only loosely substantiated by written evidence, but the possibility that it did happen greatly enhances the appeal of the earrings. The diamonds have not been removed from their settings, so exact weights are not known, but they are estimated to be about 13 and 19 carats.

Marjorie Merriweather Post bought the earrings from the jeweler Pierre Cartier in London in 1928. Cartier had bought them from Prince Felix Yousupoff, who provided a letter indicating that they had been in his family as far back as the Grand Duchess Tatiana Yousupoff, who lived from 1796 to 1841. The earrings were presented to the Smithsonian in 1964 by Eleanor Close Barzin, Mrs. Post's daughter.

Hooker Emerald

This exceptionally large 75-carat Colombian emerald set in a brooch created by Tiffany & Co., was given to the Smithsonian in 1977 by Mrs. James Stewart (Janet Annenberg) Hooker. The gem, which is unusually inclusion-free for an emerald of this size, is surrounded by 109 round diamonds, weighing a total of 10 carats, and 20 baguettes, totaling 3 carats. It is said that this magnificent emerald once adorned the belt buckle of Sultan Abdul-Hamid II, who reigned in Turkey from 1876 to 1911.

Star of Asia

This superb 330-carat royal blue star sapphire was acquired in 1961 from Martin L. Ehrmann, a California gem and mineral dealer. He accepted in exchange two lots of small faceted diamonds that had been seized by U.S. Customs and transferred to the museum. The gem is said to have belonged to the Maharajah of Jodhpur. It came from Burma (now Myanmar) and is one of the biggest and best blue star sapphires known.

Portuguese Diamond

Harry Winston was involved in the acquisition of this 127-carat gem, the museum's largest faceted diamond. The museum traded for it in 1963, giving in exchange a large lot of small diamonds considered unimportant to the collection. Nothing of the history of the diamond is known. Although it is thought to be Brazilian in origin, and to have been found in 1755, it is named for its supposed possession by the Prince Regent of Portugal. Claims of the stone's ties to Portuguese royalty are, however, unsupported.

Winston was able to trace the diamond only to its 1928 sale to Peggy Hopkins Joyce and reported that it was said to have come from South Africa and weighed 150 carats when it first arrived in New York City in 1911 or 1912. The stone's soft fluorescence under incandescent light gives it a slightly cloudy bluish look that tends to disguise its near perfection. An ultraviolet light causes it to fluoresce an intense blue-white; it can literally light up a room with its glow.

DeYoung Pink Diamond

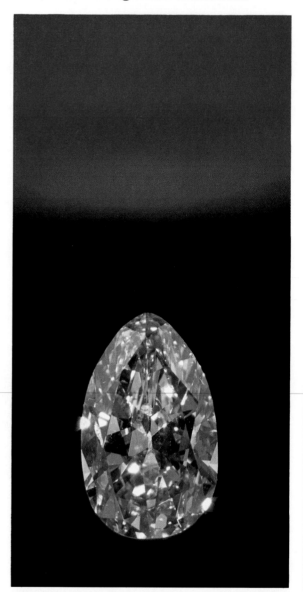

This beautiful gem diamond from Tanzania is a rare, lively pink, a more intense color than most. It was donated to the museum in 1962 by Mr. and Mrs. Sydney DeYoung. Although at only 2.9 carats it is small in comparison with most of the other important diamonds in the collection, its exceptional color never fails to delight visitors, even those who are intimately familiar with the rarer colors of diamonds.

Eugenie Blue Diamond

Not much is known of the early history of this 31-carat blue diamond. In 1960, Marjorie Merriweather Post purchased it from Harry Winston. Initially, it was called the Empress Eugenie ring, but because of confusion between it and a 51-carat white diamond named the Empress Eugenie, it became the Eugenie Blue. Mrs. Post gave it to the museum in 1964.

Logan Sapphire

One of the largest blue sapphire gems known, this stone weighs 423 carats. A fine sky blue, it is the size of a goose egg. The sapphire was mined in Sri Lanka and was donated to the museum in 1960 by Mrs. John A. (Polly) Logan, the former Rebecca Pollard Guggenheim.

Maharani Chrysoberyl Cat's-Eye

This excellent 58.2-carat chrysoberyl cat's-eye is from Sri Lanka. It is not the largest in the collection, but it is the finest and one of the best large chrysoberyl cat's-eyes in the world. The gem's outstanding features include exceptional size, fine color, and a well-defined ray of reflected light.

Index

Page numbers in **boldface** refer to illustrations.

acicular habit, **24**, 24
agate (quartz), 68–69, **69**
   dyed, **51**
alexandrite (chrysoberyl), 78, 89
amber, **84**, 84
American Museum of Natural History, 63, 71, 88
amethyst (quartz), **18**, 18, **27**, **39**, **50**, **66**, **67**, 67
ametrine (quartz), **67**, 67
ammonite, **30**, 31
apatite, **40**, 40–41
aquamarine (beryl), **15**, 53, **57**, **62**, 62
aragonite, 12, **13**
assembled stones, 87
aventurine (quartz), 68
azurite, 34

barite, **25**
barrel habit, **24**, 24
beryl, **15**, 53, **57**, **62**, 62–63, **89**, 91
beryl, aquamarine, **15**, 53, **57**, **62**, 62
beryl, bixbite, 53, **62**, 63
beryl, emerald, **63**, **89**, 91
beryl, goshenite, 63
beryl, heliodor, 53, 63
beryl, morganite, 53, **62**, 63
beryl, prisimatic, **24**
Berzelius, J.J., 38
bixbite (beryl), 53, **62**, 63
bladed habit, 24, **25**

cabochons, 56
calcite, 12, **13**, 35, **36**, 36
cameo, **85**, 85
Canfield Collection, 46
carat, 53
celestine, **25**
chalcedony (quartz), 68–69, **82**
Chamberlain, Dr. Leander T., 88, 89
charoite, **83**, 83
chert (quartz), 69
chrysoberyl, **78**, 78, 89, **93**
   alexandrite, 78, 89
   cat's-eyes, 78
   Maharani, cat's-eye, **93**
citrine (quartz), **15**, **50**, **66**, 67
Clarke, Frank W., 88
cleavage, **36**, 36, 58
color, 34–35
copper, **38**
coral, **84**, 84
cordierite, 78–79
   iolite, 78

corundum (ruby and sapphire), 39, 53, 54, 60–61, **91**, **93**
crystal habits, 24–25
crystallization, oddities of, 26–29
crystals, 21–29
crystals, twinned, **27**, 27
crystal shapes, 22–23
crystal systems, seven, 22–23, **22**
cubic zirconia, 78
cuprite crystals, electron microscope image of, **43**

Dana, James Dwight, 38
Dana system, 38–39
dendrites, 31, 31, **69**
dendritic habit, 24
diamond, 12, **32**, 53, 54, 58–59, **59**
diamond crystal, Oppenheimer, **32**
diamond, DeYoung pink, 59, **92**
diamond, DeYoung red, **58**, 89
diamond, Eugenie Blue, **93**
diamond, Hope, 89, **90**
diamond, Marie Antoinette earrings, **91**
diamond, Napoleon necklace, **59**
diamond, Pearson, **58**
diamond, Portuguese, **92**
diamond, Victoria-Transvaal, **52**
diamonds, colored, 59

earth, chemical composition of, 13
earth, subdivisions of the, **16**
elbaite (tourmaline), 76, **77**
electron microprobe, **42**, 42–43, 49
emerald, Hooker (beryl), **91**
emerald, Anna Case Mackay necklace, **63**, 89
emerald, Spanish Inquisition necklace, **89**
emerald (beryl), 53, 62, **63**, **89**
eosphorite, 29
Eugenie Blue diamond, **93**

faceted cuts, 56
faceting, 52, 56–57
fayalite, 13
feldspar, 79–80
   labradorite, **79**, 79
   moonstone, **79**, 79
   orthoclase, 79–80
   sunstone, **55**, **79**, 79
field collecting, 44–45
filiform habit, 24
flint (quartz), 69
fluorapatite, **40**, 40

fluorescence, **37**, 37
fluorite, 36, **39**, 39, **50**, 83
forsterite, 13
fossils, 31, 84
fracture, 36

galena, 36, 38
garnet, 20, 41, 72–73
   andradite, 73
   almandine, 72, **73**
   demantoid, 73
   grossular, 72, **73**, 73
   pyrope, 72, **73**
   rhodolite, 72, 73
   star, 73
   tsavorite, 73
gem weight, measuring, 53
gems, cutting, 56–59
gems, imitation, 86–87
gems, organic, 84–85
gems, pleochroic, 57
gems, "precious," 52
gems, synthetic, 61, 78, **86**, 86, 87
gems and gemstones, 15, 51–93,
geodes, **18**, 18, **26**, 26, 69
gold, 13, **38**
goshenite (beryl), 63
Grand Canyon, 13, 18, **19**
granite, 20
graphite, **12**, 12
gypsum, 21, **28**, 28, 29

halides, 39
halite, 12, **39**, 39
   crystal structure of, **22**
hardness, 32–33
heliodor (beryl), 63
hematite, 24, **25**
hiddenite (spodumene), **6**, 80
Hidden, W.E., 80
Hooker emerald, **91**
Hope diamond, 89, **90**
hydroxides, 39

igneous rocks, 16–18
inclusions, quartz, 68
iolite (cordierite), 78
ivory, 85

jade, 74–75
jadeite, **74**, 74, **75**, 75
jadeite, Dragon Vase, **75**
jasper, 69
jet, 85

94

karat, 53
Kimberley diamond mine, 54
kornerupine, 53
Kunz, George F., 63, 77, 80, 88
kunzite (spodumene), **15**, 35, **80**, 80
kyanite, 78–79

labradorite (feldspar), **79**, 79
La Brea tar pits, 31
lapidaries, 56–57
lapis lazuli, 82
Lea Collection, 88
Leidy, Dr. Joseph, 88
light, refraction of, 56
Logan sapphire, **93**
luster, 33

Maharani chrysoberyl cat's-eye, **93**
malachite, **34**, **81**, 81
Marie Antoinette earrings (diamond), **91**
metamorphic rocks, 19–20
metatyuyamunite, **23**
meteorites, 16
micromounts, **23**, 23
microscope, petrographic, 20
microscope, scanning electron, 42, **43**, 49
mimetite, **34**, 40
mineral, replacement, 30–31
mineral collecting, **44**, 44–45
minerals, chemical composition and crystalline structure, 12–13
minerals, classifying, 38–39
minerals, collecting, 44–45
minerals, isometric, 36
minerals, isotropic and anisotropic, 35
minerals, physical properties of, 32–37
minerals, radioactive, 37
mineral structures, fourteen basic, 12, **14**
Mohs, Frederick, 33
Mohs scale, 32–34, 51
moon, 10
moonstone (feldspar), **79**, 79
morganite (beryl), 53, **62**, 63
Morgan, J.P., 63, 88
mother-of-pearl, 84
Munsteiner, Bernd, 57
muscovite, **25**, 36
micaceous habit, 24, **25**

Napoleon necklace, **59**
National Gem Collection, 88–93
National Mineral Collection, 46–49
National Museum of Natural History, 7, 38, 46–49
natrolite, **24**
nephrite, **74**, 74, **75**, 75

onyx (quartz), 69
opal, 64–65
opal, black, 64
opal, Dark Jubilee, **65**
opal, fire, 64, **65**
opal, jelly, 64
opal replacing wood, **30**, 30
opal, white, **64**

organic gems, 84–85
orthoclase (feldspar), 79–80
oxides, 34, 39

parting, 36
pearls, **85**, 85
peridot, 13, 54, **76**, 76
  mining, **55**
Petrified Forest National Park, 30
petrographic microscope, 20
phantoms, crystal, **28**, 28
pleochroism, 57
plumose habit, 24, **25**
Portuguese diamond, **92**
prismatic habit, **24**, 24
pyrite, **32**, 38
  fossils, **30**, 30, 31
  "sand dollar," **31**, 31
pyromorphite, 40, **41**
pyrophyllite, **25**

quartz, **11**, 11, 12, **27**, **28**, 28, 29, 39, 66–69
quartz, agate, **51**, 68–69, **69**
quartz, amethyst, **18**, 18, **27**, 39, **50**, **66**, **67**, 67
quartz, ametrine, **67**, 67
quartz, aventurine, 68
quartz, chalcedony, 68–69, 82
quartz, chert, 69
quartz, citrine, **15**, **50**, **66**, 67
quartz, dendritic agate, **69**
quartz, flint, 69
quartz, geode, **26**, 26, **69**
quartz, onyx, 69
quartz, phantoms, **28**, 28
quartz, rock crystal, **11**, 11, **66**, 66, 67
quartz, rose, **15**, **66**, **68**, 68
quartz, scepters, **27**, 27
quartz, smoky, **66**, 67–68, **68**
quartz, thunder eggs, 69
quartz, twins, **27**, 27
quartz inclusions, 68

recrystallization, 19
refraction of light, 56
reniform habit, 24, **25**
rhodochrosite, **34**
rhodolite (garnet), 72
rock crystal (quartz), **11**, 11, **66**, 66, 67
rocks, igneous, 16–18
rocks, metamorphic, 19–20
rocks, sedimentary, 18–19
rocks and minerals, 13–20
Rocky Mountains, **9**
Roebling Collection, 46, 88
rose quartz, **15**, **66**, **68**, 68
Rosser Reeves star ruby, **61**
ruby (corundum), 60–61
ruby, star (corundum), 60–61
ruby, synthetic, 61, **86**, 86

sapphire, 54, **60**, 60–61, **61**, 93
sapphire, Logan, **93**
sapphire, star, 60–61, **91**
sapphire, Star of Asia, **91**
sapphire, Star of Bombay, 61

sapphire, synthetic, 61
scapolite, **80**, 80
scepters, crystal, **27**, 27–28
sedimentary rocks, 18–19
serpentine, 75
shell, 84–85
  helmet, **85**
silicates, 13, 39, 41
smithsonite, **47**
Smithson, James, 46
Smithson medal, **46**
smoky quartz, **66**, 67–68, **68**
Spanish Inquisition necklace (emerald), **89**
specific gravity, 36
specimen collecting, 44–45, **45**
spessartine (garnet), 72, 73
spinel, 41, **77**, 77–78
spodumene, 35, **80**, 80
  hiddenite, **6**, 80
  kunzite, **15**, 35, **80**, 80
  mine, **44**
Star of Asia sapphire, **91**
Star of Bombay sapphire, 61
star ruby, Rosser Reeves, **61**
staurolite, 20
stellate habit, 24, **25**
stibnite, **33**
sugilite, 83
sulfides, 34, 38
sulfosalts, 38
sulfur, 38
sunstone (feldspar), 55, **79**, 79

taaffeite, 53
tanzanite, **35**, 35, 80
  zoisite, **35**, 35
thunder eggs (quartz), 69
topaz, **15**, **50**, **70**, 70–71, **71**
topaz, American Golden, **40**
torbernite, **23**
tourmaline, **15**, 55, **76**, 76–77
  elbaite, **77**
  mining, **54**
tsavorite (garnet), 52, 73
turquoise, 53, 54, 81–82, **82**
twinning, **27**, 27

U.S. Geological Survey, 49

vanadinite, **24**, 40, **41**
Verneuil, Auguste, 86
volcano, Kilauea, 16, **17**

wulfenite, **23**

X-ray diffraction, 23, 42–43, 49

yttrium-aluminum-oxide crystals (YAG) (synthetic gem), **87**

zeolites, 41
zinc ore, **37**
zircon, 78
zoisite, **35**, 35
  tanzanite, 80
zoning, 28–29, **29**

ACKNOWLEDGMENTS

Minerals and Gems was conceived, developed, and produced by Gramercy Books and the Book Development Division, Smithsonian Institution Press.

Gramercy Books: Glorya Hale, Editorial Director; Frank Finamore, Assistant Editor; Helene Berinsky, Designer; Ellen Reed, Production Supervisor.

Book Development Division, Smithsonian Institution Press: Caroline Newman, Executive Editor; Paula Ballo Dailey, Picture Editor; Heidi Lumberg, Assistant Editor; Sue Voss, National Museum of Natural History, Consulting Editor.

Special thanks to Daniel E. Appleman, Merle Berk, Russell C. Feather, Paul W. Pohwat, Jeffrey E. Post, Mary T. Winter, and the Office of Printing and Photographic Services, Smithsonian Institution.

All photography by Chip Clark, National Museum of Natural History, except as noted:

Ray Albrechtsen: p. 55.
Dan Behnke: p. 23.
Richard S. Fiske: p. 45.
Foto-Studio Baumann, Hohr-Grenzhausen: p. 57.
J.D. Griggs: p. 17.
Victor Krantz: pp. 28 bottom right, 29.
Dane Penland: pp. 12 top left, 18, 32 left, 33, 38 bottom left, 47, 51, 52, 53, 58 top left, 60, 61 bottom, 66, 69 top left, 70, 71, 72, 73 center, 74, 75 top, 76 bottom left, 77 bottom, 78 left, 79 bottom, 80 center, 81, 84 right, 85 right, 89, 91 top, 91 bottom right, 92 right, 93 top and bottom right.
Jeffrey E. Post: p. 43.
Office of Printing and Photographic Services, Smithsonian Institution: pp. 46, 78 right.
U.S. Department of the Interior, National Park Service, Margaret Farrell: p. 19.
John Sampson White: pp. 44, 54.

Drawings by Clair Moritz.

Half title page: Banded agate from Chihuahua, Mexico.
Title page: Amethyst crystals from Vera Cruz, Mexico.